Finite Elements and Symmetry

Finite Elements and Symmetry

Special Issue Editor

Rachid Touzani

MDPI • Basel • Beijing • Wuhan • Barcelona • Belgrade

MDPI

Special Issue Editor
Rachid Touzani
Laboratoire de Mathématiques Blaise Pascal,
Université Clermont Auvergne
France

Editorial Office
MDPI
St. Alban-Anlage 66
4052 Basel, Switzerland

This is a reprint of articles from the Special Issue published online in the open access journal *Symmetry* (ISSN 2073-8994) from 2019 to 2020 (available at: https://www.mdpi.com/journal/symmetry/special_issues/Finite_Elements_Symmetry).

For citation purposes, cite each article independently as indicated on the article page online and as indicated below:

LastName, A.A.; LastName, B.B.; LastName, C.C. Article Title. *Journal Name* **Year**, *Article Number*, Page Range.

ISBN 978-3-03936-020-8 (Hbk)
ISBN 978-3-03936-021-5 (PDF)

Contents

About the Special Issue Editor

Rachid Touzani is full professor of Applied Mathematics at the University Clermont Auvergne (Clermont-Ferrand, France). He was born in Kénitra, Morocco in 1955. He received a M.S. degree from the Université de Franche Comté, Besançon, France, in 1979, and a Ph.D. degree from Ecole Polytechnique Fédérale de Lausanne (EPFL), Switzerland, in 1988. R. Touzani started his scientific career as a scientific researcher at the EPFL from 1982 to 1992, where he contributed to and conducted research projects in computational fluid dynamics and industrial applications involving electromagnetic processes. Since 1992, R. Touzani has continued his research on eddy current applications from both numerical and theoretical aspects. He was the head of the Mathematical modelling and engineering department at the Polytech Clermont Engineering School. R. Touzani is the author of more than 100 research articles and conferences and is coa-uthor of a textbook on mathematical modelling of eddy currents.

Preface to "Finite Elements and Symmetry"

As a numerical method for the approximation of solutions of partial differential equations, the finite element method has long since proven its efficiency, flexibility, and practicability. Specific issues in the numerical solution have been addressed using this method, such as some qualitative properties of solutions. Among these properties, positivity, regularity, and symmetry are included.

According to the research area covered by the journal *Symmetry*, this Special Issue gathered some publications relative to symmetry in finite element analysis of partial differential equations. This topic is poorly represented in the finite element literature and our objective was to compensate for this lack.

Symmetry appears under various aspects:

- Symmetries in domain geometry where this can be considered to simplify generation and adaptation of finite element meshes;
- Symmetry in boundary conditions, which can contribute to simplifingy variational formulations;
- Symmetry in the model definition, such as the use of symmetric tensors in continuum mechanics, where this property can be sought in numerical simulations; and
- Expected symmetry in solution and symmetry breaking in nonlinear bifurcation problems.

This Special Issue, entitled Finite Elements and Symmetry, aimed to collect various studies related to this topic to enrich the finite element literature from this aspect

Rachid Touzani
Special Issue Editor

symmetry

MDPI

Article

Well Posedness and Finite Element Approximability of Three-Dimensional Time-Harmonic Electromagnetic Problems Involving Rotating Axisymmetric Objects

Praveen Kalarickel Ramakrishnan † and Mirco Raffetto *,†

Department of Electrical, Electronic, Telecommunications Engineering and Naval Architecture, University of Genoa, Via Opera Pia 11a, I–16145 Genoa, Italy; pravin.nitc@gmail.com

* Correspondence: mirco.raffetto@unige.it; Tel.: +39-010-3352796

† These authors contributed equally to this work.

Received: 6 December 2019; Accepted: 19 January 2020; Published: 2 February 2020

Abstract: A set of sufficient conditions for the well posedness and the convergence of the finite element approximation of three-dimensional time-harmonic electromagnetic boundary value problems involving non-conducting rotating objects with stationary boundaries or bianisotropic media is provided for the first time to the best of authors' knowledge. It is shown that it is not difficult to check the validity of these conditions and that they hold true for broad classes of practically important problems which involve rotating or bianisotropic materials. All details of the applications of the theory are provided for electromagnetic problems involving rotating axisymmetric objects.

Keywords: electromagnetic scattering; time-harmonic electromagnetic fields; moving media; rotating axisymmetric objects; bianisotropic media; variational formulation; well posedness; finite element method; convergence of the approximation

1. Introduction

The presence of rotating objects in electromagnetic problems is of interest in several applications, ranging from the detection of helicopters to the tachometry of celestial bodies [1,2]. Unfortunately, as an immediate consequence of the presence of materials in motion, all these electromagnetic problems are difficult to solve. This is a consequence of the fact that all moving media are perceived as bianisotropic [3,4].

Independently of the motion, bianisotropic media have been considered in several recent investigations, in particular in the context of metamaterials, with frequencies belonging to the microwave band or to the photonic one [5–8], for their huge potentialities or for their practical applications.

The complexity of electromagnetic problems involving media in motion or bianisotropic materials prevents any chance of getting results without the use of numerical simulators. However, in order to rely on them, it is important to know a priori results of well posedness of the problems of interest and on their numerical approximability. A few papers addressing these topics have been recently published [9–12]. However, due to the difficulty of the problems considered, most of them present results under some restrictive hypotheses. For example, in [9], the results of interest are deduced by exploiting in a crucial way the presence of losses, while in [10] the authors study cylinders in axial motions. In [11], a problem of evolution is studied inside a cavity, preventing the exploitation of the results in many applications and, finally, in [12] the constitutive parameters are smooth so neglecting the possibility of considering radiation or scattering problems.

In this paper, we try to overcome most of these limitations by extending the theory developed in [10] to three-dimensional time-harmonic electromagnetic boundary value problems involving lossy or lossless materials which can be bianisotropic or in motion. Only on the materials in motion will we consider some restrictions. In particular, in order to retain the possibility to perform the analysis of time-harmonic problems, we need that the boundaries of the moving objects are stationary [3]. Thus, we will restrict ourselves to consider the rotation of axisymmetric objects. For the same reason, the velocity field will be considered independent of time. Moreover, the media in motion have to be non-conductive, in order to avoid the difficulties related to the convective currents, which could become surface electric currents [13] and then determine a discontinuity of the tangential part of the magnetic field.

As for the media involved whose bianisotropy is not due to motion, we do not consider any restrictive hypothesis. In particular, the formulation we consider allows the solution of radiation [14], scattering [1,5,15], or guided wave problems [16,17], which are all of interest for applications.

The well posedness and finite element approximability guaranteed by our theory allow us to obtain reliable solutions from numerical simulations for rotating axisymmetric objects. With this, we can solve several problems. However, for the sake of conciseness, we selected just two representative examples. For one of them, we have approximate semi-analytic solutions [1], and the range of validity of the approximation involved in those solutions can be verified using our approach. Our second example is representative of the majority of problems involving rotating objects, for which no result can be found in the open literature. For any problem of this class, the reliable solution obtained under the conditions required by our theory can serve as a benchmark for other numerical techniques.

The paper is organized as follows. In Section 2, the problems of interest are defined. Section 3 reports the main ideas which can be used to show that the problems of interest are well posed. The results of convergence of Galerkin and finite element approximations are presented in Section 4. In Section 5, we briefly present the main features of the finite element simulator exploited to compute the results presented in Section 7. In these first sections, we heavily exploit the results presented in [9,10,18]. We have included these sections in our manuscript in order to ease readers' task and because the results we present are not trivially deduced from [10,18], since they deal with two-dimensional problems. The main novelties of the paper are presented in Sections 6 and 7. In particular, in Section 6, we present some useful suggestions on how our theory can be exploited to solve problems of practical interest and in Section 7 the practical applications of our theory to rotating axisymmetric objects are presented. The conclusions are reported in Section 8 and some technical details are provided in the appendix.

2. Problem Definition

In this section, we define the time-harmonic electromagnetic boundary value problem we will deal with in the rest of the paper. Most of the considerations of this section are taken from Sections 2 and 3 of [9] and are here reported to ease the reader's task and to introduce some specific considerations of interest for problems involving rotating axisymmetric objects.

To avoid restrictions on the applicability of our analysis, the problem will be formulated on a domain Ω satisfying the following hypotheses ($\Gamma = \partial\Omega$ denotes its boundary):

HD1. $\Omega \subset \mathbb{R}^3$ is open, bounded and connected,
HD2. Γ is Lipschitz continuous and stationary.

Moreover, in order to be able to consider electromagnetic problems of practical interest, different inhomogeneous materials will be taken into account. This is the reason why we assume:

HD3. Ω can be decomposed into m subdomains (non-empty, open and connected subsets of Ω having Lipschitz continuous stationary boundaries) denoted Ω_i, $i \in I = \{1, \ldots, m\}$, satisfying $\overline{\Omega} = \overline{\Omega}_1 \cup \ldots \cup \overline{\Omega}_m$ ($\overline{\Omega}$ is the closure of Ω) and $\Omega_i \cap \Omega_j = \emptyset$ for $i \neq j$.

This hypothesis allows us to consider also the presence of rotating axisymmetric objects.

The specific target of the paper is to deal with electromagnetic problems involving very general materials. However, in order to give a sense to a time-harmonic analysis, we have at least to assume that:

HM1. Any material involved is linear and time-invariant and satisfies the following constitutive relations:

$$\begin{cases} \mathbf{D} = (1/c_0)\, P\, \mathbf{E} + L\, \mathbf{B} & \text{in } \Omega, \\ \mathbf{H} = M\, \mathbf{E} + c_0\, Q\, \mathbf{B} & \text{in } \Omega. \end{cases} \tag{1}$$

In the above equation, $\mathbf{E}$, $\mathbf{B}$, $\mathbf{D}$, $\mathbf{H}$, and c_0 are, respectively, the electric field, the magnetic induction, the electric displacement, the magnetic field and the velocity of light in vacuum [19]. L, M, P and Q are four 3-by-3 matrix-valued complex functions defined almost everywhere in Ω. The vector fields $\mathbf{E}$, $\mathbf{B}$, $\mathbf{D}$ and $\mathbf{H}$ are complex valued too, as it is usually the case for electromagnetic field problems in which the real fields depend sinusoidally on time [20] (pp. 13–16). Equation (1) implicitly takes account of the electric current densities, as usual. Other equivalent forms of the above constitutive equations are possible [21] (p. 49) [22], and will also be used later on.

Different inhomogeneous bianisotropic materials will be modeled by assuming the following hypothesis.

HM2. The matrix valued complex functions representing the effective constitutive parameters satisfy [23] (p. 3), [24] (p. 36):

$$P|_{\Omega_k}, Q|_{\Omega_k}, L|_{\Omega_k}, M|_{\Omega_k} \in (C^0(\overline{\Omega}_k))^{3\times 3}, \; \forall k \in I$$

Such hypothesis is in no way restrictive for all applications of interest since the material properties are just piecewise but not globally continuous. In particular, as we will verify later on, hypotheses HM1 and HM2 do not exclude the presence of rotating axisymmetric objects [2] either.

The following additional notations and hypotheses are necessary too. $(L^2(\Omega))^3$ is the usual Hilbert space of complex-valued square integrable vector fields on Ω and with scalar product given by $(\mathbf{u}, \mathbf{v})_{0,\Omega} = \int_\Omega \mathbf{v}^* \mathbf{u}\, dV$ (* denotes the conjugate transpose). $H(\text{curl}, \Omega) = \{\mathbf{v} \in (L^2(\Omega))^3 \mid \text{curl}\, \mathbf{v} \in (L(\Omega))^3\}$ [24] (p. 55). The space where we will seek $\mathbf{E}$ and $\mathbf{H}$ is [24] (p. 82; see also p. 69)

$$U = H_{L^2,\Gamma}(\text{curl}, \Omega) = \{\mathbf{v} \in H(\text{curl}, \Omega) \mid \mathbf{v} \times \mathbf{n} \in L_t^2(\Gamma)\}, \tag{2}$$

where [24] (p. 48)

$$L_t^2(\Gamma) = \{\mathbf{v} \in (L^2(\Gamma))^3 \mid \mathbf{v} \cdot \mathbf{n} = 0 \;\text{ almost everywhere on } \Gamma\}. \tag{3}$$

The scalar products in $L_t^2(\Gamma)$ and U are respectively given by $(\mathbf{u}, \mathbf{v})_{0,\Gamma} = \int_\Gamma \mathbf{v}^* \mathbf{u}\, dS$ and [24] (p. 84, p. 69)

$$(\mathbf{u}, \mathbf{v})_{U,\Omega} = (\mathbf{u}, \mathbf{v})_{0,\Omega} + (\text{curl}\, \mathbf{u}, \text{curl}\, \mathbf{v})_{0,\Omega} + (\mathbf{u} \times \mathbf{n}, \mathbf{v} \times \mathbf{n})_{0,\Gamma}. \tag{4}$$

The induced norm is $\|\mathbf{u}\|_U = (\mathbf{u}, \mathbf{u})_{U,\Omega}^{1/2}$.

The symbol ω represents the angular frequency, as usual. Moreover, $\mathbf{J}_e$ and $\mathbf{J}_m$ are the electric and magnetic current densities, respectively, prescribed by the sources, Y is the scalar admittance involved in impedance boundary condition and $\mathbf{f}_R$ is the corresponding inhomogeneous term. Finally, the admittance function Y with domain Γ and range in $\mathbb{C}$ is assumed to satisfy

HB1. Y is piecewise continuous and $|Y|$ is bounded.

We are now in a position to state the electromagnetic boundary value problem we will address in this paper.

Problem 1. *Under the hypotheses HD1-HD3, HM1-HM2, HB1, given* $\omega > 0$, $\mathbf{J}_e \in (L^2(\Omega))^3$, $\mathbf{J}_m \in (L^2(\Omega))^3$ *and* $\mathbf{f}_R \in L_t^2(\Gamma)$, *find* $(\mathbf{E}, \mathbf{B}, \mathbf{H}, \mathbf{D}) \in U \times (L^2(\Omega))^3 \times U \times (L^2(\Omega))^3$ *satisfying (1) and the following equations:*

$$\begin{cases} \operatorname{curl} \mathbf{H} - j\omega\, \mathbf{D} = \mathbf{J}_e & \text{in } \Omega, \\ \operatorname{curl} \mathbf{E} + j\omega\, \mathbf{B} = -\mathbf{J}_m & \text{in } \Omega, \\ \mathbf{H} \times \mathbf{n} - Y\left(\mathbf{n} \times \mathbf{E} \times \mathbf{n}\right) = \mathbf{f}_R & \text{on } \Gamma. \end{cases} \tag{5}$$

As it was pointed out in [9], such a model can be thought of as an approximation of a radiation or scattering problem, or as a realistic formulation of a cavity problem.

The following variational formulation of Problem 1 was derived in [9]:

Problem 2. *Under the hypotheses HD1–HD3, HM1–HM2, HB1, given* $\omega > 0$, $\mathbf{J}_e \in (L^2(\Omega))^3$, $\mathbf{J}_m \in (L^2(\Omega)^3$ *and* $\mathbf{f}_R \in L_t^2(\Gamma)$, *find* $\mathbf{E} \in U$ *such that*

$$a(\mathbf{E}, \mathbf{v}) = l(\mathbf{v}) \ \ \forall \mathbf{v} \in U, \tag{6}$$

where

$$\begin{aligned} a(\mathbf{u}, \mathbf{v}) = c_0\left(Q \operatorname{curl} \mathbf{u}, \operatorname{curl} \mathbf{v}\right)_{0,\Omega} - \frac{\omega^2}{c_0}\left(P\, \mathbf{u}, \mathbf{v}\right)_{0,\Omega} - j\omega\left(M\, \mathbf{u}, \operatorname{curl} \mathbf{v}\right)_{0,\Omega} \\ -j\omega\left(L \operatorname{curl} \mathbf{u}, \mathbf{v}\right)_{0,\Omega} + j\omega\left(Y\left(\mathbf{n} \times \mathbf{u} \times \mathbf{n}\right), \mathbf{n} \times \mathbf{v} \times \mathbf{n}\right)_{0,\Gamma} \end{aligned} \tag{7}$$

and

$$l(\mathbf{v}) = -j\omega\left(\mathbf{J}_e, \mathbf{v}\right)_{0,\Omega} - c_0\left(Q\, \mathbf{J}_m, \operatorname{curl} \mathbf{v}\right)_{0,\Omega} + j\omega\left(L\, \mathbf{J}_m, \mathbf{v}\right)_{0,\Omega} - j\omega\left(\mathbf{f}_R, \mathbf{n} \times \mathbf{v} \times \mathbf{n}\right)_{0,\Gamma}. \tag{8}$$

It was shown in [9] that the two formulations are equivalent, in the sense that, from the solution of Problem 1, one can deduce the solution of Problem 2 and vice versa; moreover, the well posedness of the former implies the well posedness of the latter and vice versa [9].

3. Well Posedness of the Problem

Following the main ideas presented in Section 4 of [10], in this section, we prove the well posedness of the three-dimensional problems of interest. The target will be achieved by showing that, under appropriate additional hypotheses, we can apply the generalized Lax-Milgram lemma [24] (p. 21) to Problem 2.

The continuity of the sesquilinear and antilinear forms, a and l, are easily deduced under the hypotheses already introduced (HD1-HD3, HM1-HM2, HB1). Thus, it remains to introduce the additional hypotheses allowing us to prove that the sesquilinear form a satisfies the following conditions:

$$\text{for every } \mathbf{v} \in U, \mathbf{v} \neq 0, \quad \sup_{\mathbf{u} \in U} |a(\mathbf{u}, \mathbf{v})| > 0, \tag{9}$$

$$\text{we can find } \alpha : \inf_{\mathbf{u} \in U,\, \|\mathbf{u}\|_U = 1} \ \sup_{\mathbf{v} \in U,\, \|\mathbf{v}\|_U \leq 1} |a(\mathbf{u}, \mathbf{v})| \geq \alpha > 0. \tag{10}$$

We establish under which hypotheses these conditions hold true in the following subsections.

3.1. Hypotheses to Prove Condition (9)

Condition (9) is easily proved once we know that the solution to Problem 2 is unique, as shown in [10]. In turn, uniqueness for Problem 2 is achieved by proving uniqueness for the corresponding homogeneous problem (that is the one with $l = 0$) [25] (p. 20), [24] (p. 92). Finally, uniqueness for the corresponding homogeneous problem can be deduced by a standard technique [26] (pp. 187–203), [10,24,27] (p. 92), in the presence of some losses and by unique continuation results.

In the following, we introduce the hypotheses which allow for getting the result of interest in this subsection. In order to let the reader understand the general picture, we observe that:

- the first group of hypotheses (HM3 and HB2) requires that the media and the boundary do not provide active power,
- the second group of assumptions (HM4–HM7 and HB3) asks for the presence of some losses in the media or on the boundary or the invertibility of the constitutive matrix P, $\forall \mathbf{x} \in \overline{\Omega}_i, \forall i \in I$,
- the first two groups of hypotheses are sufficient to prove that the solution of the homogeneous problem is zero on a subdomain of Ω or that its tangential part on a subset of the boundary is zero,
- the third group of assumptions (HM8–HM12) guarantee the applicability of a unique continuation result, allowing us to show that condition (9) holds true.

In order to write our assumptions, we need to introduce some additional notation.

In [9], it was shown that the sesquilinear form a can be recast is the form

$$a(\mathbf{u},\mathbf{v}) = \int_{\Omega} \left\{ (\mathbf{v}^*, \operatorname{curl} \mathbf{v}^*) \, A \begin{pmatrix} \mathbf{u} \\ \operatorname{curl} \mathbf{u} \end{pmatrix} \right\} + j\omega \left(Y \mathbf{n} \times \mathbf{u} \times \mathbf{n}, \mathbf{n} \times \mathbf{v} \times \mathbf{n} \right)_{0,\Gamma'} \tag{11}$$

where

$$A = \begin{pmatrix} -\frac{\omega^2}{c_0} P & -j\omega L \\ -j\omega M & c_0 Q \end{pmatrix} = A_s - jA_{ss}, \tag{12}$$

being [9] $A_s = \frac{A+A^*}{2}$ and $A_{ss} = \frac{A^*-A}{2j}$. For future use, the vector notation introduced in Equation (11) is generalized as follows for the ordered pair $\mathbf{q}, \mathbf{r} \in \mathbb{C}^3$:

$$\mathbf{p} = \begin{pmatrix} \mathbf{q} \\ \mathbf{r} \end{pmatrix}. \tag{13}$$

Moreover, by referring to the constitutive relation (1) or the above definition of A, we introduce a splitting of the subscript $i \in I$ of the subdomains Ω_i: $i \in I_a$ when $L = M = 0 \; \forall \mathbf{x} \in \Omega_i$ (the media are anisotropic), otherwise $i \in I_b$. Finally, an alternative form of the constitutive relations will be used to state unique continuation results. Such an alternative form is

$$\begin{cases} \mathbf{E} = \kappa \, \mathbf{D} + \chi \, \mathbf{B} & \text{in } \Omega, \\ \mathbf{H} = \gamma \, \mathbf{D} + \nu \, \mathbf{B} & \text{in } \Omega, \end{cases} \tag{14}$$

where the constitutive matrices $\kappa = c_0 \, P^{-1}$, $\chi = -c_0 \, P^{-1} \, L$, $\gamma = c_0 \, M \, P^{-1}$ and $\nu = c_0 \, (Q - M \, P^{-1} \, L)$ [22] are all well defined where P^{-1} is well defined (see hypothesis HM7 below).

The first group of hypotheses is the following:

HM3. $\mathbf{p}^* A_{ss} \mathbf{p} \leq 0$, $\forall \mathbf{p} \in \mathbb{C}^6$, $\forall \mathbf{x} \in \Omega_i$, $\forall i \in I$,

HB2. $Re(Y) \geq 0$ on Γ.

The assumptions of the second group (HM4–HM7 on the media and HB3 on the boundary) are all related to the presence of losses (apart from HM7) and read:

HM4. We can find $K_{dl} > 0$ and $D \subset \Omega_i$, $i \in I$, D open, non-empty such that $\mathbf{p}^* A_{ss} \mathbf{p} \leq -K_{dl}(|\mathbf{q}|^2 + |\mathbf{r}|^2)$ in D,

HM5. We can find $K_{el} > 0$ and $D \subset \Omega_i$, $i \in I$, D open, non-empty such that $\mathbf{p}^* A_{ss} \mathbf{p} \leq -K_{el} |\mathbf{q}|^2$ in D,

HM6. We can find $K_{ml} > 0$ and $D \subset \Omega_i$, $i \in I_a$, D open, non-empty such that $\mathbf{p}^* A_{ss} \mathbf{p} \leq -K_{ml} |\mathbf{r}|^2$ in D,

HM7. P is invertible, for all $\mathbf{x} \in \overline{\Omega}_i$, $\forall i \in I$,

HB3. We can find $C_{Ym} > 0$ and a non-empty open part Γ_l of Γ such that $Re(Y) \geq C_{Ym}$ almost everywhere on Γ_l.

Appropriate combinations of these hypotheses are sufficient to prove (see Lemma A1 in the Appendix A) that any solution of the homogeneous variational problem has a tangential part, which is trivial on Γ_l or is trivial in the subdomain D.

Once this result has been obtained, in order to prove that the field is zero everywhere in Ω, one has to apply unique continuation results [26] (pp. 187–203), [10,24,27] (p. 92). To achieve this target in the presence of anisotropic and bianisotropic media, we refer to [22], and introduce the following third set of hypotheses:

HM8. All entries of $\kappa, \chi, \gamma, \nu \in C^{\infty}(\overline{\Omega}_i)$ and are restrictions of analytic functions in Ω_i, $\forall i \in I$,

HM9. $\exists\exists C_{\kappa,d} > 0,\ C_{\nu,d} > 0 : |determinant\,(\kappa)\,| \geq C_{\kappa,d},\ |determinant\,(\nu)\,| \geq C_{\nu,d},\ \forall \mathbf{x} \in \overline{\Omega}_i, \forall i \in I$,

HM10. $\mathbf{l}_{1,3}^T\,\kappa^{-1}\,\mathbf{l}_{1,3} \neq 0,\ \mathbf{l}_{1,3}^T\,\nu^{-1}\,\mathbf{l}_{1,3} \neq 0\ \forall \mathbf{l}_{1,3} \in \mathbb{R}^3, \mathbf{l}_{1,3} \neq 0,\ \forall \mathbf{x} \in \overline{\Omega}_i,\ \forall i \in I_a$,

HM11. $\exists\exists C_{\kappa,r} > 0,\ C_{\nu,r} > 0 :\ |\mathbf{l}_{1,3,n}^T\,\kappa^{-1}\,\mathbf{l}_{1,3,n}| \geq C_{\kappa,r},\ |\mathbf{l}_{1,3,n}^T\,\nu^{-1}\,\mathbf{l}_{1,3,n}| \geq C_{\nu,r}\ \forall \mathbf{l}_{1,3,n} \in \mathbb{R}^3 : \|\mathbf{l}_{1,3,n}\|_2 = 1,\ \forall \mathbf{x} \in \overline{\Omega}_i,\ \forall i \in I_b$,

HM12. $\exists\exists C_{\kappa,s} > 0, C_{\nu,s} > 0$:

$$\Big(\sum_{i,j=1}^{3} |\kappa_{ij}|\Big) - \min_{i=1,2,3} |\kappa_{ii}| \leq C_{\kappa,s} \quad \forall \mathbf{x} \in \overline{\Omega}_k,\ \forall k \in I_b, \tag{15}$$

$$\Big(\sum_{i,j=1}^{3} |\nu_{ij}|\Big) - \min_{i=1,2,3} |\nu_{ii}| \leq C_{\nu,s} \quad \forall \mathbf{x} \in \overline{\Omega}_k,\ \forall k \in I_b, \tag{16}$$

and κ, χ, γ and ν satisfy

$$\frac{4\left(\left(\sum_{i,j=1}^{3} |\gamma_{ij}|\right) - \min_{i=1,2,3} |\gamma_{ii}|\right)\left(\left(\sum_{i,j=1}^{3} |\chi_{ij}|\right) - \min_{i=1,2,3} |\chi_{ii}|\right)}{\left(-C_{\kappa,s} + \sqrt{C_{\kappa,s}^2 + 4\,C_{\kappa,d}\,C_{\kappa,r}}\right)\left(-C_{\nu,s} + \sqrt{C_{\nu,s}^2 + 4\,C_{\nu,d}\,C_{\nu,r}}\right)} < 1 \tag{17}$$

$\forall \mathbf{x} \in \overline{\Omega}_k,\ \forall k \in I_b$.

Remark 1. *The constants and the constraints involved in hypotheses HM9, HM11 and HM12 could be defined in any single subdomain Ω_i, $i \in I_b$, in order to deduce less restrictive conditions under which our theory holds true. This approach was exploited for example in [10]. Here, we use constants and constraints defined globally, in order to avoid longer and technically more complicated definitions.*

In particular, with hypotheses HM7, HM8, HM9 and HM10, by Theorem 6.4 of [22], we can conclude that any solution of the homogeneous variational problem is analytic in all anisotropic media, i.e., for all Ω_i, $i \in I_a$. Moreover, under hypotheses HM7, HM8, HM9, HM11 and HM12, by Theorem 7.3 of [22], we get the same result for all Ω_i, $i \in I_b$.

These preliminary outcomes allow us to state the following uniqueness result, which will be proved in Appendix A:

Theorem 1. *Under the hypotheses HD1–HD3, HM1–HM3, HM7–HM9, HB1–HB2, if HM10 is satisfied by the anisotropic media and HM11 and HM12 are satisfied by the bianisotropic materials involved, then Problem 2 admits a unique solution provided that at least one of HM4 or HM5 or HM6 or HB3 is satisfied.*

Like in [10,28], it is now extremely simple to deduce (in Appendix A, it is possible to find the proof; * denotes the complex conjugate)

Theorem 2. *Under the hypotheses HD1–HD3, HM1–HM3, HM7–HM9, HB1–HB2, if HM10 is satisfied by the anisotropic media and HM11 and HM12 are satisfied by the bianisotropic materials involved, then the homogeneous variational problem, find* $\mathbf{v} \in U$ *such that* $(a(\mathbf{u},\mathbf{v}))^* = 0\ \forall \mathbf{u} \in U$*, admits a unique solution* $\mathbf{v} = 0$ *provided that at least one of HM4 or HM5 or HM6 or HB3 is satisfied.*

With this result, we can finally show that, under appropriate hypotheses, condition (9) holds true.

Theorem 3. *Under the hypotheses HD1–HD3, HM1–HM3, HM7–HM9, HB1–HB2, if HM10 is satisfied by the anisotropic media and HM11 and HM12 are satisfied by the bianisotropic materials involved, then condition (9) holds true provided that at least one of HM4 or HM5 or HM6 or HB3 is satisfied.*

Proof. Suppose that (9) is not satisfied. Then, we can find $\mathbf{v} \in U, \mathbf{v} \neq 0$ such that $\sup_{\mathbf{u} \in U} |a(\mathbf{u},\mathbf{v})| = 0$. However, $|a(\mathbf{u},\mathbf{v})| = |\ (a(\mathbf{u},\mathbf{v}))^*\ |$. Then, for the indicated $\mathbf{v} \neq 0$, $|\ (a(\mathbf{u},\mathbf{v}))^*\ | = 0\ \forall \mathbf{u} \in U$. This is at odds with Theorem 2, since we have assumed the same hypotheses. □

3.2. Additional Hypotheses to Prove Condition (10)

Under hypothesis HM2 or HB1, by a direct application of the Cauchy–Schwarz inequality, we deduce that it is possible to define the following continuity constants:

- $\exists C_{PL} > 0$: $|(P\,\mathbf{u},\mathbf{v})_{0,\Omega}| \leq C_{PL}\|\mathbf{u}\|_{0,\Omega}\|\mathbf{v}\|_{0,\Omega}$ for all $\mathbf{u},\mathbf{v} \in (L^2(\Omega))^3$,
- $\exists C_L > 0$: $|(L\,\text{curl}\,\mathbf{u},\mathbf{v})_{0,\Omega}| \leq C_L\|\text{curl}\,\mathbf{u}\|_{0,\Omega}\|\mathbf{v}\|_{0,\Omega}$ for all $\mathbf{u} \in H(\text{curl},\Omega)$ and $\mathbf{v} \in (L^2(\Omega))^3$,
- $\exists C_M > 0$: $|(M\,\mathbf{u},\text{curl}\,\mathbf{v})_{0,\Omega}| \leq C_M\|\mathbf{u}\|_{0,\Omega}\|\text{curl}\,\mathbf{v}\|_{0,\Omega}$ for all $\mathbf{u} \in (L^2(\Omega))^3$ and $\mathbf{v} \in H(\text{curl},\Omega)$,
- $\exists C_{YL} > 0$: $|(Y(\mathbf{n}\times\mathbf{u}\times\mathbf{n}),\mathbf{n}\times\mathbf{v}\times\mathbf{n})_{0,\Gamma}| \leq C_{YL}\|\mathbf{n}\times\mathbf{u}\times\mathbf{n}\|_{0,\Gamma}\|\mathbf{n}\times\mathbf{v}\times\mathbf{n}\|_{0,\Gamma}$

In order to prove condition (10), we introduce the following additional hypotheses, which guarantee that it is possible to find some coercivity constants:

HM13. We can find $C_{PS} > 0$ such that $|(P\mathbf{u},\mathbf{u})_{0,\Omega}| \geq C_{PS}\|\mathbf{u}\|^2_{0,\Omega}$ for all $\mathbf{u} \in (L^2(\Omega))^3$.

HM14. We can find $C_{QS} > 0$ such that $|(Q\text{curl}\,\mathbf{u},\text{curl}\,\mathbf{u})_{0,\Omega}| \geq C_{QS}\|\text{curl}\,\mathbf{u}\|^2_{0,\Omega}$ for all $\mathbf{u} \in H(\text{curl},\Omega)$.

HB3S. We can find $C_{Ym} > 0$ such that $Re(Y) \geq C_{Ym}$ almost everywhere on Γ.

Moreover, we assume:

HM15. C_{PS}, C_{QS}, C_L and C_M (i.e., all media involved) are such that $C_{QS} - \frac{C_L C_M}{C_{PS}} > 0$.

As is shown in Appendix A, it is now possible to get the following result:

Theorem 4. *Under the hypotheses HD1–HD3, HM1–HM3, HM7–HM9, HB1, HB3S, HM13–HM15, if HM10 is satisfied by the anisotropic media and HM11 and HM12 are satisfied by the bianisotropic materials involved, then the sesquilinear form a satisfies condition (10).*

The following theorem, which is the main result of this section, is now a simple consequence:

Theorem 5. *Under the hypotheses HD1–HD3, HM1–HM3, HM7–HM9, HB1, HB3S, HM13–HM15, if HM10 is satisfied by the anisotropic media and HM11 and HM12 are satisfied by the bianisotropic materials involved, then Problem 2 is well posed.*

Proof. HB3S implies HB2 and HB3. It also implies that the logical or of HM4, HM5, HM6 and HB3, which is present as a condition in Theorem 3, is true. Thus, the hypotheses reported in the statement of the theorem guarantee the applicability of Theorems 3 and 4. □

4. Convergence of Galerkin and Finite Element Approximations

Once the result of well posedness of the problems of interest is established, we can proceed as in Sections 5 and 6 of [10], to deduce the conditions under which the convergence of Galerkin [29] and finite element [24] approximations can be guaranteed.

Convergence of an approximation [29] (p. 112) refers to the property of sequences of solutions of the approximate problem and requires that they converge to the unique solution of the problem of interest.

Any sequence of approximate solutions is built by considering a sequence $\{U_h\}$ of finite dimensional subspaces U_h of U. h is a denumerable and bounded set of strictly positive indexes having zero as the only limit point [29] (p. 112).

For any $h \in I$, a set of approximate sources is considered: $\mathbf{J}_{eh}, \mathbf{J}_{mh} \in (L^2(\Omega))^3$ and $\mathbf{f}_{Rh} \in L^2_t(\Gamma)$. With these, we define the following approximate antilinear form:

$$l_h(\mathbf{v}) = -j\omega(\mathbf{J}_{eh}, \mathbf{v})_{0,\Omega} - c_0(Q\,\mathbf{J}_{mh}, \mathrm{curl}\mathbf{v})_{0,\Omega} + j\omega(L\,\mathbf{J}_{mh}, \mathbf{v})_{0,\Omega} - j\omega(\mathbf{f}_{Rh}, \mathbf{n} \times \mathbf{v} \times \mathbf{n})_{0,\Gamma} \quad (18)$$

and the following discrete version of Problem 2.

Problem 3. *Under the hypotheses HD1–HD3, HM1–HM2, HB1, given* $\omega > 0$, $\mathbf{J}_{eh} \in (L^2(\Omega))^3$, $\mathbf{J}_{mh} \in (L^2(\Omega)^3$ *and* $\mathbf{f}_{Rh} \in L^2_t(\Gamma)$, *find* $\mathbf{E}_h \in U_h$ *such that*

$$a(\mathbf{E}_h, \mathbf{v}_h) = l_h(\mathbf{v}_h) \;\; \forall \mathbf{v}_h \in U_h. \quad (19)$$

In order to state the results of interest, it is necessary to introduce the following subspaces of U_h:

$$U_{0h} = \{\mathbf{u}_h \in U_h \,|\, \mathrm{curl}\,\mathbf{u}_h = 0 \text{ in } \Omega \text{ and } \mathbf{u}_h \times \mathbf{n} = 0 \text{ on } \Gamma\}, \quad (20)$$

$$U_{1h} = \{\mathbf{u}_h \in U_h \,|\, (P\mathbf{u}_h, \mathbf{v}_h)_{0,\Omega} = 0 \,\forall \mathbf{v}_h \in U_{0h}\}. \quad (21)$$

On the sequence of approximating space [24,30], we need to consider

HSAS1. $\lim_{h\to 0} \inf_{\mathbf{u}_h \in U_h} \|\mathbf{u} - \mathbf{u}_h\|_U = 0, \; \forall \mathbf{u} \in U,$

HSAS2. from any subsequence $\{\mathbf{u}_{h1}\}_{h\in I}$ of elements $\mathbf{u}_{h1} \in U_{1h}$ which is bounded in U, one can extract a subsequence converging strongly in $(L^2(\Omega))^3$ to an element of U,

HSAS3. $\lim_{h\to 0} \inf_{\mathbf{u}_{0h} \in U_{0h}} \|\mathbf{u}_0 - \mathbf{u}_{0h}\|_U = 0.$

To get meaningful approximations, the sequences of discrete sources have to satisfy:

HSDS1. $\lim h \to 0 \|\mathbf{J}_e - \mathbf{J}_{eh}\|_{0,\Omega} = 0,$

HSDS2. $\lim h \to 0 \|\mathbf{J}_m - \mathbf{J}_{mh}\|_{0,\Omega} = 0,$

HSDS3. $\lim h \to 0 \|\mathbf{f}_R - \mathbf{f}_{Rh}\|_{0,\Gamma} = 0.$

The following is one of the main results of this section:

Theorem 6. *Under the hypotheses HD1–HD3, HM1–HM3, HM7–HM9, HB1, HB3S, HM13–HM15, HSAS1–HSAS3, HSDS1–HSDS3, if HM10 is satisfied by the anisotropic media and HM11 and HM12 are satisfied by the bianisotropic materials involved, then the sequence* $\{\mathbf{E}_h\}$ *of solutions of Problem 3 strongly converges to* $\mathbf{E} \in U$, $\mathbf{E}$ *being the unique solution of Problem 2.*

Proof. The proof is only sketched being analogous to that of Theorem 5.3 of [10]. The first part of the proof shows that, under the indicated hypotheses, for any sufficiently small $h \in I$, we get a unique solution $\mathbf{E}_h$ of Problem 3.

Thus, since the hypotheses guarantee also the well posedness of Problem 2, we can deal, for sufficiently small $h \in I$, with $\mathbf{E}$ and $\mathbf{E}_h$.

The last part of the proof verifies that the sequence $\|\mathbf{E} - \mathbf{E}_h\|_U$ strongly converges to zero. □

The sequence of finite dimensional subspaces for the Galerkin approximation is typically built using the finite element method [29]. This involves the use of a sequence of triangulations $\{\mathcal{T}_h\}$, $h \in I$, of $\overline{\Omega}$ and a specific finite element on each triangulation $\mathcal{T}_h$ [29].

To avoid some technicalities arising with curved boundaries, we assume that [29] (p. 65)

HD4. Ω is a polyhedron (i.e., $\overline{\Omega} = \bigcup_{T \in \mathcal{T}_h} T$).

Edge elements defined on tetrahedra are very often employed for approximating fields belonging to $H(\text{curl}, \Omega)$. For this reason, we assume [29–31]:

HFE1. the family $\{\mathcal{T}_h\}$ of triangulations is regular,

HFE2. $\mathcal{T}_h$ is made up of tetrahedra, $\forall h \in I$,

HFE3. edge elements of a given order defined on tetrahedra are used to build U_h, $\forall h \in I$.

By classical results in finite element theory, we can now conclude that whenever HD1, HD2, HD4, HFE1–HFE3 are satisfied, the space sequence $\{U_h\}$ verifies conditions HSAS1, HSAS2 and HSAS3.

Thus, we obtain the second main results of this section:

Theorem 7. *Under the hypotheses HD1–HD4, HM1–HM3, HM7–HM9, HB1, HB3S, HM13–HM15, HSDS1–HSDS3, HFE1–HFE3, if HM10 is satisfied by the anisotropic media and HM11 and HM12 are satisfied by the bianisotropic materials involved, then Problem 3 is a convergent approximation of Problem 2.*

5. Some Information about the Exploited Finite Element Simulator

In this section, we provide some specific considerations related to the implementation of our finite element code that was used to obtain the numerical solutions to the problems. A first order edge element based Galerkin approach is adopted [32], and most of the details are analogous to the two-dimensional implementation found in [18]. For any mesh adopted, we get the finite dimensional space U_h. In it, we can find the test functions $\mathbf{v}_{hi}$, $i \in \{1, ..., ne\}$, where ne is the number of edges of the mesh. Then, denoting the vector of unknowns as $[e_h] \in \mathbb{C}^{ne}$ and using Equations (7), (18) and (19), we can obtain the following matrix equation:

$$[A_h][e_h] = [l_h]. \tag{22}$$

Here, $[A_h]$ is the complex matrix whose entries are obtained from Equation (7) and are given by:

$$\begin{aligned} [A_h]_{ij} &= c_0 \left(Q \,\text{curl}\, \mathbf{v}_{hj}, \text{curl}\, \mathbf{v}_{hi}\right)_{0,\Omega} - \frac{\omega^2}{c_0} \left(P\, \mathbf{v}_{hj}, \mathbf{v}_{hi}\right)_{0,\Omega} - j\omega \left(M\, \mathbf{v}_{hj}, \text{curl}\, \mathbf{v}_{hi}\right)_{0,\Omega} \\ &- j\omega \left(L \,\text{curl}\, \mathbf{v}_{hj}, \mathbf{v}_{hi}\right)_{0,\Omega} + j\omega \left(Y\left(\mathbf{n} \times \mathbf{v}_{hj} \times \mathbf{n}\right), \mathbf{n} \times \mathbf{v}_{hi} \times \mathbf{n}\right)_{0,\Gamma}, \; i, j = 1, ..., ne. \end{aligned} \tag{23}$$

The entries $[l_h]_i$ are obtained trivially from (18) by replacing $\mathbf{v}$ with $\mathbf{v}_{hi}$. In general, $[A_h]$ is a non-Hermitian complex matrix and in our approach we made use of iterative methods for the solution of the algebraic system. In particular, we exploited the biconjugate gradient method with Jacobi preconditioner [33]. The solution $[e_h]_i$ obtained in the i-th iteration is accepted only when the Euclidean norm of error satisfies $||[A_h][e_h]_i - [l_h]|| < \delta ||[l_h]||$. Here, δ is a fixed value denoting the acceptable tolerance, which is set as $\delta = 10^{-p}$, p being an integer (see Section 5 of [18,33]). For the test problems of Sections 7.3 and 7.4, the value p was set equal to 10 and 5, respectively. The solutions obtained were checked for convergence by refining the mesh until stable results were achieved.

6. Some Hints to Apply the Developed Theory

The developed theory required the introduction of 32 hypotheses: four on the domain (HD1–HD4), four on the boundary conditions (HB1–HB3 and HB3S), 15 on the media involved (and, as it will be shown in Section 7, on the way, they rotate; HM1–HM15), three on the sequence of approximating space (HSAS1–HSAS3), three on the sequence of discrete sources (HSDS1–HSDS3) and three on the finite element discretization (HFE1–HFE3).

The main results of this manuscript, related to the well posedness of the problem of interest and to the convergence of its finite element approximation, make use, respectively, of 17 and 24 of these assumptions.

In order to ease the exploitation of the main outcomes, we observe that most of these hypotheses can be verified immediately for important practical problems. This is true, in particular, for conditions HD1–HD4, HB1–HB3 and HB3S, HM1–HM8, HSDS1–HSDS3, and HFE1–HFE3. Hypotheses HSAS1–HSAS3 are not involved in the indicated theorems. As for the other hypotheses to be verified, in the following, we provide some hints which can be of help to show that assumptions HM9–HM15 holds true.

Let us firstly focus on the additional hypotheses we have introduced to prove condition (10) (that is, HM13 and HM14). In this section, we extensively use the notation introduced in Equation (12) and the line following it.

One simpler way to find the constant involved in hypothesis HM13 is provided by the following Lemma.

Lemma 1. *Suppose that P_{ss} is uniformly positive definite in $\Omega_{el} \subset \Omega$ that is $\exists C_1 > 0$ such that*

$$\int_{\Omega_{el}} \mathbf{u}^* P_{ss} \mathbf{u} \geq C_1 \int_{\Omega el} |\mathbf{u}|^2 = C_1 ||\mathbf{u}||^2_{0,\Omega_{el}} \; \forall \mathbf{u} \in (L^2(\Omega))^3. \tag{24}$$

Whenever $\Omega_{el} = \Omega$, we can simply define $C_{PS} = C_1$.

Whenever Ω_{el} is not the whole Ω, suppose that, in the complementary region, P_s is uniformly positive or negative definite, that is, $\exists C_5 > 0$ such that

$$\left| \int_{\Omega \backslash \Omega_{el}} \mathbf{u}^* P_s \mathbf{u} \right| \geq C_5 ||\mathbf{u}||^2_{0,\Omega \backslash \Omega_{el}}. \tag{25}$$

Whenever $\Omega_{el} = \emptyset$, we simply have $C_{PS} = C_5$ and we can set

$$C_{PS} = \min_{i \in I} \inf_{\mathbf{x} \in \Omega_i} \lambda_{min}(P_s), \tag{26}$$

where λ_{min} denotes the minimum of the magnitudes of the eigenvalues of the Hermitian symmetric matrix P_s.

Finally, whenever Ω_{el} is neither the empty set nor the whole domain, under assumptions HM2 and HM3, condition HM13 is satisfied with C_{PS} given by

$$C_{PS} = \frac{1}{\sqrt{2}} \min \left(\sqrt{(1-\alpha)} C_5, \sqrt{C_1^2 + (1 - \frac{1}{\alpha}) C_3^2} \right), \tag{27}$$

where $C_3 > 0$ is defined by

$$\left| \int_{\Omega_{el}} \mathbf{u}^* P_s \mathbf{u} \right| \leq C_3 ||\mathbf{u}||^2_{0,\Omega_{el}} \tag{28}$$

and α is such that $1 > \alpha > \frac{C_3^2}{C_1^2 + C_3^2} > 0$.

Lemma 1 is proved in the Appendix A by using a technique developed in [34].

In an analogous way, by replacing P with Q in Equations (24), (25) and (28) , we define, respectively, Ω_{ml} and the constants $C_2 > 0$, $C_4 > 0$ and $C_6 > 0$ and deduce that condition HM14 is satisfied if we set

$$C_{QS} = \min_{i \in I} \inf_{\mathbf{x} \in \Omega_i} \lambda_{min}(Q_s), \tag{29}$$

whenever $\Omega_{ml} = \emptyset$, $C_{QS} = C_2$ whenever $\Omega_{ml} = \Omega$ or

$$C_{QS} = \frac{1}{\sqrt{2}} \min \left(\sqrt{(1-\alpha)C_6}, \sqrt{C_2^2 + (1 - \frac{1}{\alpha})C_4^2} \right), \tag{30}$$

being α such that $1 > \alpha > \frac{C_4^2}{C_2^2 + C_4^2} > 0$, when $\Omega_{ml} \neq \Omega$ and $\Omega_{ml} \neq \emptyset$.

The above lemma will be heavily exploited to show the applicability of our theory to many practical problems of interest. However, it does not imply that it is not possible to find larger values of C_{PS}. For example, whenever P_s is uniformly definite in Ω that is $\exists C_7 > 0$ such that

$$\left| \int_\Omega \mathbf{u}^* P_s \mathbf{u} \right| \geq C_7 ||\mathbf{u}||_{0,\Omega}^2, \tag{31}$$

we can choose for C_{PS} the largest between C_7 and the value obtained by using Lemma 1.

This is of interest in order to reduce the restrictions due to inequality HM15. In order to check its validity, we also have to evaluate the continuity constants $C_L > 0$ and $C_M > 0$. From their very definitions, one can estimate these values and set for example

$$C_L = \max_{i \in I_b} \sup_{\mathbf{x} \in \Omega_i} \sqrt{\lambda_{max}(L^* L)} \tag{32}$$

and

$$C_M = \max_{i \in I_b} \sup_{\mathbf{x} \in \Omega_i} \sqrt{\lambda_{max}(M^* M)}, \tag{33}$$

where λ_{max} denotes the maximum of the magnitudes of the eigenvalues of the Hermitian symmetric matrix to which it applies.

We now look for simple techniques to check the validity of hypotheses HM9–HM12. Our previous considerations assume that we know the constitutive matrices P, Q, L and M. The next ones, on the contrary, are based on κ, ν, χ and γ. In order to deduce this form of the constitutive parameters, one can use the equations reported below Equation (14) under hypothesis HM7.

To check the validity of assumptions HM9–HM12, the constants $C_{\kappa,d}$, $C_{\nu,d}$, $C_{\kappa,r}$, $C_{\nu,r}$, $C_{\kappa,s}$ and $C_{\nu,s}$ have to be evaluated (see Remark 1). For $C_{\kappa,d}$, $C_{\nu,d}$, $C_{\kappa,s}$ and $C_{\nu,s}$ one has simply to apply the definitions, for example by calculating

$$C_{\kappa,d} = \min_{i \in I} \inf_{\mathbf{x} \in \Omega_i} |determinant(\kappa)|, \tag{34}$$

$$C_{\nu,d} = \min_{i \in I} \inf_{\mathbf{x} \in \Omega_i} |determinant(\nu)|, \tag{35}$$

$$C_{\kappa,s} = \max_{i \in I} \sup_{\mathbf{x} \in \Omega_i} \left((\sum_{i,j=1}^{3} |\kappa_{ij}|) - \min_{i=1,2,3} |\kappa_{ii}| \right), \tag{36}$$

$$C_{\nu,s} = \max_{i \in I} \sup_{\mathbf{x} \in \Omega_i} \left((\sum_{i,j=1}^{3} |\nu_{ij}|) - \min_{i=1,2,3} |\nu_{ii}| \right). \tag{37}$$

As for $C_{\kappa,r}$ and $C_{\nu,r}$ the following consideration might be helpful. By definition

$$C_{\kappa,r} = \min_{i \in I} \inf_{\mathbf{x} \in \Omega_i} \min_{\mathbf{1}_{1,3,n} \in \mathbb{R}^3 : \|\mathbf{1}_{1,3,n}\|_2 = 1} \sqrt{\left(\mathbf{1}_{1,3,n}^T \kappa_{is} \mathbf{1}_{1,3,n}\right)^2 + \left(\mathbf{1}_{1,3,n}^T \kappa_{iss} \mathbf{1}_{1,3,n}\right)^2}, \tag{38}$$

$$C_{\nu,r} = \min_{i\in I} \inf_{\mathbf{x}\in\Omega_i} \min_{\mathbf{l}_{1,3,n}\in\mathbb{R}^3:\|\mathbf{l}_{1,3,n}\|_2=1} \sqrt{\left(\mathbf{l}_{1,3,n}^T \nu_{is} \mathbf{l}_{1,3,n}\right)^2 + \left(\mathbf{l}_{1,3,n}^T \nu_{iss} \mathbf{l}_{1,3,n}\right)^2}, \tag{39}$$

where κ_{is} and κ_{iss} are the symmetric matrices obtained by the usual decomposition of κ^{-1} and similarly ν_{is} and ν_{iss} are those corresponding to ν^{-1}. If both the symmetric matrices involved in the above expressions are semi-definite, then we can deduce the following lower bounds:

$$C_{\kappa,r} = \min_{i\in I} \inf_{\mathbf{x}\in\Omega_i} \sqrt{(\lambda_{min}(\kappa_{is}))^2 + (\lambda_{min}(\kappa_{iss}))^2}, \tag{40}$$

$$C_{\nu,r} = \min_{i\in I} \inf_{\mathbf{x}\in\Omega_i} \sqrt{(\lambda_{min}(\nu_{is}))^2 + (\lambda_{min}(\nu_{iss}))^2}. \tag{41}$$

If we also define

$$C_{\chi,s} = \max_{i\in I} \sup_{\mathbf{x}\in\Omega_i} \left((\sum_{i,j=1}^{3} |\chi_{ij}|) - \min_{i=1,2,3}|\chi_{ii}| \right), \tag{42}$$

$$C_{\gamma,s} = \max_{i\in I} \sup_{\mathbf{x}\in\Omega_i} \left((\sum_{i,j=1}^{3} |\gamma_{ij}|) - \min_{i=1,2,3}|\gamma_{ii}| \right), \tag{43}$$

the sufficient condition for the regularity used for proving uniqueness can be expressed as

$$K_u = \frac{4C_{\chi,s}C_{\gamma,s}}{\left(-C_{\kappa,s} + \sqrt{C_{\kappa,s}^2 + 4C_{\kappa,d}C_{\kappa,r}}\right)\left(-C_{\nu,s} + \sqrt{C_{\nu,s}^2 + 4C_{\nu,d}C_{\nu,r}}\right)} < 1. \tag{44}$$

7. Implications for Rotating Axisymmetric Objects

In this section, we show the implications of the developed theory for three-dimensional problems involving rotating axisymmetric objects.

The class of scattering problems of interest involves rotating axisymmetric objects illuminated by time-harmonic electromagnetic fields. Even though our theory does not limit the number of objects involved, in this section, we show the results computed in the presence of just one rotating rigid body (with angular velocity ω_s) because, on the one hand, this is enough to get bianisotropic effects and, on the other hand, notwithstanding the limitation, it is still possible to define problems whose solutions, to the best of the authors' knowledge, is not known. In these cases, our solutions may then be considered as benchmarks.

By the same token, it is not necessary to consider very complicated configurations of materials. This is the reason why in this subsection we analyze problems involving objects rotating in vacuum. In our notation, the empty space is characterized by $P = c_0\varepsilon_0 I_3$, $Q = \frac{1}{c_0\mu_0} I_3$, $L = M = 0$, being I_3 the identity matrix. In order to avoid problems with convective currents, which can become surface currents [35], we assume that all rotating media in their rest frames have the electric conductivity $\sigma = 0$ and real-valued ε and μ. However, we need to know the constitutive parameters when the media are rotating. To get these results, we recall that for media in motion with a generic velocity field $\mathbf{v}$ we have [19] (p. 958)

$$\mathbf{D} + \frac{1}{c_0^2}\mathbf{v}\times\mathbf{H} = \varepsilon\left(\mathbf{E} + \mathbf{v}\times\mathbf{B}\right), \tag{45}$$

$$\mathbf{B} - \frac{1}{c_0^2}\mathbf{v}\times\mathbf{E} = \mu\left(\mathbf{H} - \mathbf{v}\times\mathbf{D}\right). \tag{46}$$

If $\mu \neq 0$, from Equation (46), one immediately gets

$$\mathbf{H} = \frac{1}{\mu}\mathbf{B} - \frac{1}{\mu c_0^2}\left(\mathbf{v}\times\mathbf{E}\right) + \left(\mathbf{v}\times\mathbf{D}\right), \tag{47}$$

and, by substituting it in Equation (45), one easily deduces

$$\mathbf{D} - \frac{1}{c_0^2}\left(\mathbf{v} \times \mathbf{D}\right) \times \mathbf{v} = \varepsilon \mathbf{E} - \frac{1}{\mu c_0^4}\left(\mathbf{v} \times \mathbf{E}\right) \times \mathbf{v} + \frac{\mu_r \varepsilon_r - 1}{\mu c_0^2}\left(\mathbf{v} \times \mathbf{B}\right). \tag{48}$$

Cross multiplying (on the left) this equation by $\frac{\mathbf{v}}{v^2}$, being $v = |\mathbf{v}|$, one obtains

$$\left(\mathbf{v} \times \mathbf{D}\right) = \frac{\mu_r \varepsilon_r c_0^2 - v^2}{\mu c_0^2 (c_0^2 - v^2)}\left(\mathbf{v} \times \mathbf{E}\right) - \frac{1}{\mu}\frac{\mu_r \varepsilon_r - 1}{c_0^2 - v^2}\left(\mathbf{v} \times \mathbf{B}\right) \times \mathbf{v} \tag{49}$$

and, by substituting it in the expression of $\mathbf{H}$, one gets [36]

$$\mathbf{H} = \frac{\mu_r \varepsilon_r - 1}{\mu (c_0^2 - v^2)}\left(\mathbf{v} \times \mathbf{E}\right) + \frac{1}{\mu}\mathbf{B} - \frac{\mu_r \varepsilon_r - 1}{\mu (c_0^2 - v^2)}\left(\mathbf{v} \times \mathbf{B}\right) \times \mathbf{v}. \tag{50}$$

Finally, if one obtains $(\mathbf{v} \times \mathbf{D}) \times \mathbf{v}$ from Equation (49) and substitutes the result in Equation (48), the following expression is obtained:

$$\mathbf{D} = \varepsilon \mathbf{E} + \frac{\mu_r \varepsilon_r - 1}{\mu c_0^2 (c_0^2 - v^2)}\left(\mathbf{v} \times \mathbf{E}\right) \times \mathbf{v} + \frac{\mu_r \varepsilon_r - 1}{\mu (c_0^2 - v^2)}\left(\mathbf{v} \times \mathbf{B}\right). \tag{51}$$

The last two equations allow us to find the constitutive parameters of the rotating media as perceived in the laboratory frame. Without loss of generality, we can assume that z is the axis of rotation of the rigid body. Then, the velocity field is along the azimuthal direction and has a magnitude given by the constant angular velocity ω_s multiplied by the distance of the considered point from the z axis. In the chosen Cartesian reference frame, one immediately gets $\mathbf{v} = \omega_s(x\hat{\mathbf{y}} - y\hat{\mathbf{x}})$. Then, for a generic vector $\mathbf{A}$, one deduces $\mathbf{v} \times \mathbf{A} = \omega_s x A_z \hat{\mathbf{x}} + \omega_s y A_z \hat{\mathbf{y}} - \omega_s (x A_x + y A_y)\hat{\mathbf{z}}$ and $(\mathbf{v} \times \mathbf{A}) \times \mathbf{v} = \omega_s^2 (x^2 A_x + xy A_y)\hat{\mathbf{x}} + \omega_s^2 (xy A_x + y^2 A_y)\hat{\mathbf{y}} + \omega_s^2 A_z (x^2 + y^2)\hat{\mathbf{z}}$. By using these expressions in Equations (50) and (51), after simple calculations, one finds the following explicit expressions of the constitutive matrices P, Q, L and M [36]

$$P = a_1 I_3 + b_1 T_1, \tag{52}$$

$$Q = a_2 I_3 - b_1 T_1, \tag{53}$$

$$L = M = \frac{c_0 b_1}{\omega_s} T_2, \tag{54}$$

where $a_1 = \varepsilon_0 \varepsilon_r c_0$, $a_2 = \frac{1}{\mu_0 \mu_r c_0}$, b_1 is the field $\frac{\omega_s^2 (\epsilon_r \mu_r - 1)}{\mu_0 \mu_r c_0 (c_0^2 - \omega_s^2 (x^2 + y^2))}$,

$$T_1 = \begin{bmatrix} x^2 & xy & 0 \\ xy & y^2 & 0 \\ 0 & 0 & x^2 + y^2 \end{bmatrix}, \tag{55}$$

and

$$T_2 = \begin{bmatrix} 0 & 0 & x \\ 0 & 0 & y \\ -x & -y & 0 \end{bmatrix}. \tag{56}$$

Now, we may apply the theory developed in the previous sections to check when these problems are well posed.

7.1. Checking Condition (9) for Problems Involving Rotating Objects

Rotating objects are of particular interest for scattering problems. For this class of problems, it is usual to have absorbing boundary conditions, so that HB3S is satisfied in any case.

To verify conditions HM9–HM12, we calculate κ, χ, γ and ν of the scatterer by using the equations reported below Equation (14). We get:

$$\kappa = c_0 P^{-1} = \frac{c_0}{a_1 + b_1(x^2+y^2)} I + \frac{c_0}{a_1 + b_1(x^2+y^2)} \frac{b_1}{a_1} [(x^2+y^2)I - T_1], \tag{57}$$

$$\chi = -\frac{c_0}{a_1 + b_1(x^2+y^2)} \frac{c_0 b_1}{\omega_s} T_2, \tag{58}$$

$$\gamma = -\chi, \tag{59}$$

$$\nu = a_1 a_2 \kappa. \tag{60}$$

Now, we proceed as indicated in the second part of Section 6 (the one relative to the check of conditions HM9–HM12). In particular, we start calculating the determinant of κ and ν in the scatterer

$$determinant(\kappa) = \frac{c_0^3}{a_1(a_1 + b_1(x^2+y^2))^2}, \tag{61}$$

$$determinant(\nu) = \frac{a_1^3 a_2^3 c_0^3}{a_1(a_1 + b_1(x^2+y^2))^2}. \tag{62}$$

Since in vacuum $\kappa = \frac{1}{\varepsilon_0} I$ and $\nu = \frac{1}{\mu_0} I$, the above determinants reduces respectively to $\frac{1}{\varepsilon_0^3}$ and $\frac{1}{\mu_0^3}$. In order to simplify the analysis and consider the most interesting cases, we restrict our analysis to scatterers made up of homogeneous non-magnetic materials ($\mu_r = 1$) having $\varepsilon_r > 1$. Under this condition in the scatterer, we have $b_1 > 0$ and then $a_1(a_1 + b_1(x^2+y^2))^2 > a_1^3$, so that $determinant(\kappa) < \frac{c_0^3}{a_1^3} = \frac{1}{\varepsilon_0^3 \varepsilon_r^3} < \frac{1}{\varepsilon_0^3}$ and $determinant(\nu) < a_2^3 c_0^3 = \frac{1}{\mu_0^3}$. Thus, by using Equations (34) and (35), the constants $C_{\kappa,d}$ and $C_{\nu,d}$ can be determined by finding the smallest values of the determinants in the scatterer, which is found when the field $b_1(x^2+y^2)$ gets its largest value. Since $b_1(x^2+y^2)$ is an increasing function of x^2+y^2, we finally get

$$C_{\kappa,d} = \frac{c_0^3}{a_1(a_1 + b_{1,max} R^2)^2} \tag{63}$$

and

$$C_{\nu,d} = \frac{c_0^3 a_1^2 a_2^3}{(a_1 + b_{1,max} R^2)^2}, \tag{64}$$

where R is the largest distance of the boundary of the scatterer from its axis of rotation and $b_{1,max}$ is the value which the field b_1 gets for this value of x^2+y^2:

$$b_{1,max} = \frac{\omega_s^2(\varepsilon_r - 1)}{\mu_0 c_0 (c_0^2 - \omega_s^2 R^2)}. \tag{65}$$

For problems involving objects in motion, it is usual practice to introduce the maximum normalized velocity $\beta = \frac{\omega_s R}{c_0} < 1$. In terms of β, we get $b_{1,max} R^2 = \frac{(\varepsilon_r - 1)\beta^2}{\mu_0 c_0 (1-\beta^2)}$ and then

$$C_{\kappa,d} = \frac{(1-\beta^2)^2}{\varepsilon_0^3 \varepsilon_r (\varepsilon_r - \beta^2)^2} \tag{66}$$

and

$$C_{\nu,d} = \frac{\varepsilon_r^2(1-\beta^2)^2}{\mu_0^3(\varepsilon_r-\beta^2)^2}. \tag{67}$$

If now we look for the constants $C_{\kappa,r}$ and $C_{\nu,r}$, we observe that $\kappa^{-1} = \frac{1}{c_0}P$ everywhere while $\nu^{-1} = (a_1a_2\kappa)^{-1} = \frac{1}{a_1a_2c_0}P$ in the scatterer and $\nu^{-1} = \mu_0 I$ in vacuum. Moreover, P is a real symmetric positive definite matrix, both inside and outside the scatterer, and we can use Equations (40) and (41) with $\kappa_{iss} = 0$ and $\nu_{iss} = 0$. Finally, the eigenvalues of P are a_1 and $a_1 + b_1(x^2+y^2)$ in the scatterer and $c_0\varepsilon_0$ in vacuum. Thus, the minimum of the infimum of the λ_{min} involved in those expressions is achieved in both cases in vacuum and we get

$$C_{\kappa,r} = \varepsilon_0, \tag{68}$$

and

$$C_{\nu,r} = \mu_0. \tag{69}$$

Moreover, $C_{\kappa,s}$ can be deduced by computing the suprema reported in Equation (36), inside and outside the scatterer. After some calculation, one can find that inside the scatterer the supremum is equal to $\frac{2}{\varepsilon_0\varepsilon_r}$ and outside it is $\frac{2}{\varepsilon_0}$, so that

$$C_{\kappa,s} = \frac{2}{\varepsilon_0}. \tag{70}$$

In an analogous way, we get

$$C_{\nu,s} = \frac{2}{\mu_0}. \tag{71}$$

Finally, by using Equations (58) and (59), we get that the suprema reported in Equations (42) and (43) are equal to zero outside of the scatterer and strictly positive inside it. After a few calculations, we get such strictly positive quantities

$$C_{\gamma,s} = C_{\chi,s} = \frac{2\sqrt{2}c_0^2 b_{1,max} R}{\omega_s(a_1 + b_{1,max}R^2)} = \frac{2\sqrt{2}c_0^2(\varepsilon_r-1)\beta}{\varepsilon_r-\beta^2}. \tag{72}$$

Now, to satisfy condition (9), we can substitute the previous expressions of $C_{\kappa,d}$, $C_{\nu,d}$, $C_{\kappa,r}$, $C_{\nu,r}$, $C_{\kappa,s}$, $C_{\nu,s}$, $C_{\gamma,s}$ and $C_{\chi,s}$. We get

$$\begin{aligned} 1 > K_u &= \frac{4C_{\chi,s}C_{\gamma,s}}{\left(-C_{\kappa,s} + \sqrt{C_{\kappa,s}^2 + 4C_{\kappa,d}C_{\kappa,r}}\right)\left(-C_{\nu,s} + \sqrt{C_{\nu,s}^2 + 4C_{\nu,d}C_{\nu,r}}\right)} = \\ &= \frac{32\varepsilon_r(\varepsilon_r-1)^2\beta^2}{\left(-2\varepsilon_r(\varepsilon_r-\beta^2) + 2\sqrt{\varepsilon_r^4 + \varepsilon_r + \beta^4\varepsilon_r(\varepsilon_r+1) - 2\beta^2\varepsilon_r(\varepsilon_r^2+1)}\right)} \cdot \\ &\cdot \frac{1}{\left(-2(\varepsilon_r-\beta^2) + 2\sqrt{2\varepsilon_r^2 + \beta^4(\varepsilon_r^2+1) - 2\beta^2\varepsilon_r(\varepsilon_r+1)}\right)}. \end{aligned} \tag{73}$$

In Figure 1, K_u is plotted with respect to β, with ε_r as a parameter. It shows that the range $[0, \beta_{\text{critical}}]$ of β for which the validity of condition (9) is guaranteed becomes larger and larger as ε_r gets smaller and smaller, as expected. However, our analysis provides quantitative results on such a range. As it is easy to check, it is so large that no significant restriction on β emerges for practical applications.

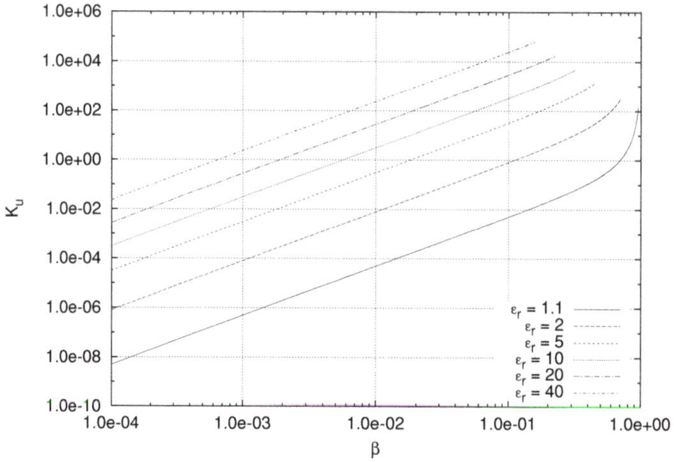

Figure 1. Plot of K_u versus β for rotating axisymmetric objects. The plots are shown for various values of ε_r. Condition (9) is satisfied for $K_u < 1$.

The plot of β_{critical} is shown, together with another significant threshold value obtained in the next subsection, in Figure 2.

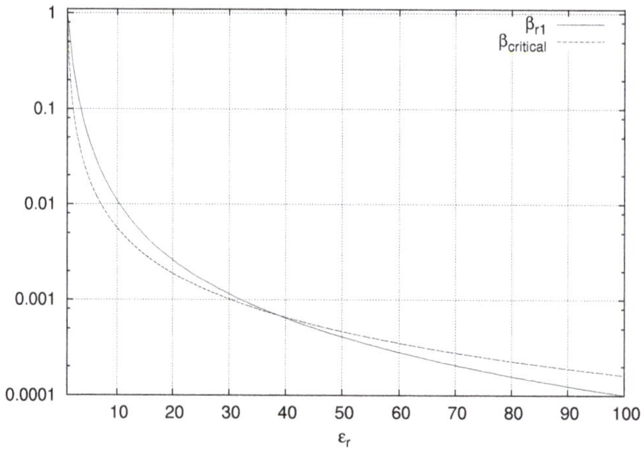

Figure 2. Behaviours of β_{r1} and β_{critical} versus ε_r. β_{critical} is the upper bound on β required to satisfy condition (9) while β_{r1} is that required for condition (10).

7.2. Checking Conditions (10) for Problems Involving Rotating Objects

In this section, we examine the situations in which condition (10) holds true for the class of problems considered. By definition, inside and outside the scatterer, we get $P_s = P$, $P_{ss} = 0$, $Q_s = Q$, $Q_{ss} = 0$. In order to check the indicated condition, we need to find the constants C_{PS}, C_{QS}, C_L and C_M. As for C_L and C_M, by using Equations (32) and (33), we have to evaluate the suprema involved just inside the scatterer. Since $M = L$, we can focus just on one of the two constants. The eigenvalues of L^*L are found to be 0 and $\left(\frac{c_0 b_1}{\omega_s}\right)^2(x^2 + y^2)$ (with multiplicity 2). As already pointed out, in the following, in order to simplify the analysis, we assume that the scatterer medium is characterized by $\varepsilon_r > 1$ and $\mu_r = 1$ in its rest frame. Under this hypothesis, the field b_1 is strictly positive and then

$$C_L = C_M = b_{1,max}\frac{c_0 R}{\omega_s}. \tag{74}$$

We already know that inside the scatterer the eigenvalues of P_s are $a_1 = \varepsilon_0\varepsilon_r c_0$ and $a_1 + b_1(x^2 + y^2)$ while outside it we have $P_s = c_0\varepsilon_0 I_3$. Under the indicated hypotheses for the scatterer medium, since $\Omega_{el} = \emptyset$, from Lemma 1 (see Equation (26)), we trivially get that HM13 is satisfied with

$$C_{PS} = \varepsilon_0 c_0. \tag{75}$$

Similarly, the eigenvalues of Q_s inside the scatterer are $a_2 = \frac{1}{c_0\mu_0\mu_r} = \frac{1}{c_0\mu_0}$ and $a_2 - b_1(x^2 + y^2)$ while outside the rotating object we have $Q_s = \frac{1}{c_0\mu_0} I_3$. Since $\Omega_{ml} = \emptyset$ by Equation (29), we obtain

$$C_{QS} = a_2 - b_{1,max}R^2, \tag{76}$$

which is positive when $\beta < \frac{1}{\sqrt{\varepsilon_r}}$. Under this condition, HM14 is satisfied as well.

By using Equations (74)–(76), the crucial inequality which is present in assumption HM15 reads

$$C_{QS} - \frac{C_L C_M}{C_{PS}} = a_2 - b_{1,max}R^2 - b_{1,max}^2\frac{c_0 R^2}{\varepsilon_0\omega_s^2} > 0. \tag{77}$$

After the substitution of a_2 and $b_{1,max}$, it can be shown to be equivalent to the following:

$$1 + \beta^2(\varepsilon_r - 2 + \varepsilon_r^2) + \beta^4\varepsilon_r > 0. \tag{78}$$

The left-hand side in inequality (78) is a parabola in terms of β^2. We can find two roots β_{r1}^2, β_{r2}^2 given by

$$\begin{cases} \beta_{r1}^2 = \frac{\varepsilon_r^2+2-\varepsilon_r-\sqrt{(\varepsilon_r^2+2-\varepsilon_r)^2-4\varepsilon_r}}{2\varepsilon_r} = \frac{\varepsilon_r^2+2-\varepsilon_r-(\varepsilon_r-1)\sqrt{\varepsilon_r^2+4}}{2\varepsilon_r} \\ \beta_{r2}^2 = \frac{\varepsilon_r^2+2-\varepsilon_r+\sqrt{(\varepsilon_r^2+2-\varepsilon_r)^2-4\varepsilon_r}}{2\varepsilon_r} = \frac{\varepsilon_r^2+2-\varepsilon_r+(\varepsilon_r-1)\sqrt{\varepsilon_r^2+4}}{2\varepsilon_r} \end{cases} \tag{79}$$

which are both real numbers. Such numbers are positive because the parabola becomes larger and larger for $\beta \to \infty$ and is equal to 1 and has a negative derivative (equal to $\varepsilon_r - 2 - \varepsilon_r^2$) when $\beta = 0$.

In particular, we have that

$$\beta_{r1}^2 < \frac{1}{\varepsilon_r} < 1, \tag{80}$$

since $\beta_{r1}^2 - \frac{1}{\varepsilon_r} = \frac{(\varepsilon_r-1)(\varepsilon_r-\sqrt{\varepsilon_r^2+4})}{2\varepsilon_r} < 0$ and $\beta_{r2}^2 > 1$ because $\beta_{r2}^2 - 1 = \frac{(\varepsilon_r-1)(\varepsilon_r-2+\sqrt{\varepsilon_r^2+4})}{2\varepsilon_r} > 0$.

Since a value greater than one is not possible for β, condition HM15 can only be satisfied for β in the range $[0, \beta_{r1}]$. In the same range of β condition, HM14 is a priori satisfied (see Equation (80) and the comment after Equation (76)) and then (10) does hold true.

The behaviours of β_{r1} and β_{critical} versus ε_r are shown in Figure 2. In order to satisfy conditions (9) and (10) and then to obtain the well posedness of the problem, β should be smaller than the smallest of β_{r1} and β_{critical}. The two plots in Figure 2 cross at about $\varepsilon_r \simeq 38.5$ and for smaller (respectively, larger) values the stronger condition on β is given by condition (9) (respectively, (10)).

7.3. Application to Rotating Sphere

In this subsection, we apply the theory to a specific case: a rotating sphere of radius R_s is illuminated by a linearly polarized plane wave propagating along the x-axis. A first order approximation of the solution of this problem is given by the semi-analytic procedure discussed by De Zutter in [1].

Our formulation of the problem requires the definition of a bounded domain Ω, which is taken as a sphere of radius R_d. The boundary conditions we enforce on Γ have Y equal to the admittance of vacuum and are inhomogeneous ($\mathbf{f}_R \neq 0$), to take account of the incident field.

The parameters considered are $\varepsilon_r = 8$, $\mu_r = 1$, $R_s = 1$ m, $R_d = 4$ m. The incident plane wave has a frequency of 50 MHz and an amplitude of the electric field of 1 V/m.

In order to analyze significant test cases for our theory and, at the same time, show its generality, we consider exceptionally large rotational speeds, without worrying about the mechanical stability of the rigid body. The rotating speed we consider is $\omega_s = 8.0\ 10^{-3} c_0$ rad/s, which corresponds to a maximum normalized velocity of $\beta = 8.0\ 10^{-3}$. This is within the limits of applicability of our theory since for $\varepsilon_r = 8$ we get $\beta_{r1} = 1.728 \times 10^{-2}$ and $\beta_{critical} = 8.124 \times 10^{-3}$. The above qualitative considerations, which apply also to the next test case (see Subsection 7.4), justify the simplified approach we have adopted (see Remark 1).

A comparison of the first order edge element based Galerkin finite element solution against the De Zutter semi-analytic procedure is carried out when the incident field is polarized along the z-axis and the spherical domain is discretized rather uniformly with a mesh having 475,797 nodes and 2,496,192 tetrahedra.

All values of the significant quantities defining our model are reported in Table 1. It includes also the parameters defining the model considered in the next subsection.

Table 1. Values of the parameters defining our models.

Type of Problem	Radius of the Domain	Geometrical Parameters of the Scatterers	Incident Plane Wave	Scatterers Constitutive Parameters	Maxima of the Normalized Velocity	Mesh of Tetrahedra
rotating sphere	4 m	$R_s = 1$ m	$f = 50$ MHz, $\lvert\mathbf{E}\rvert = 1$ V/m, propagation axis: x, polarization: linear, z	$\sigma = 0$, $\varepsilon_r = 8$, $\mu_r = 1$	$\beta = 8\ 10^{-3}$	475,797 nodes, 2,496,192 elements
rotating torus	2 m	$R = 0.15$ m, $r = 0.15$ m	$f = 500$ MHz, $\lvert\mathbf{E}\rvert = 1$ V/m, propagation axis: x, polarization: linear, z	$\sigma = 0$, $\varepsilon_r = 20$, $\mu_r = 1$	$\beta = 1.8\ 10^{-3}$	36,993 nodes, 2,192,940 elements

Figures 3 and 4 show, respectively, the magnitude and phase of the components of the electric field evaluated along a circle in the xz plane, which is centered at the origin and has a radius of 1.5 m. The results obtained from the finite element solver are compared with the semi-analytical solution obtained using the De Zutter procedure [1]. All three components are in very good agreement. Due to the well posedness and the finite element approximability of the problem, this shows that the De Zutter first order (in β) approximation provides reliable results even for very large rotational speeds.

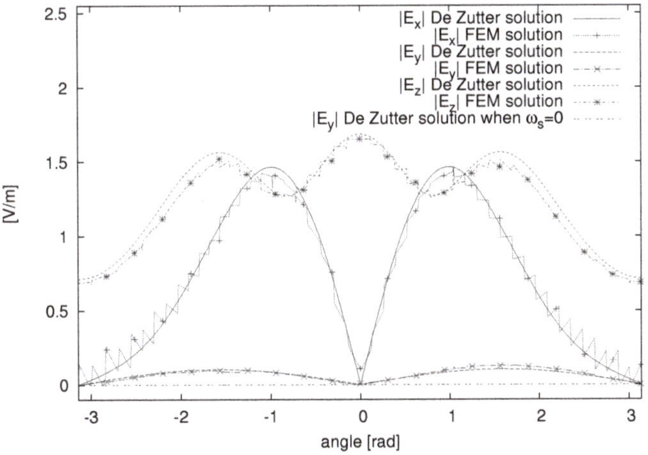

Figure 3. Comparison of the magnitudes of the electric field components along a circle in the xz plane at 1.5 m from the center of the rotating sphere. The horizontal axis represents the angle measured in radians from the x-axis.

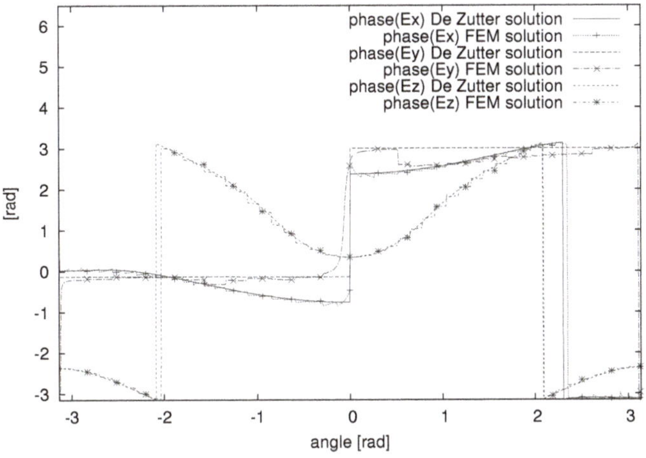

Figure 4. Comparison of the phases of the electric field components along a circle in xz plane at 1.5 m from the center of the rotating sphere. The horizontal axis represents the angle measured in radians from the x-axis.

In particular, we can observe that the y-component of the field is purely a result of rotation. This component amounts to 10 percent of the incident field. These kinds of effects on the fields can be particularly important for inverse problems to figure out the rotational speeds, for example, by extending the algorithms discussed in [37,38].

The same sort of agreement between the two solutions is further confirmed by the fields along similar circles on other planes or along lines parallel to coordinate axes, for different polarizations and directions of propagation of the illuminating field.

For example, Figures 5 and 6 show the magnitude and phase of the z component of electric field along the y-axis. Along this line, the motion of the sphere causes a difference in magnitude of up to 20 percent of the incident field.

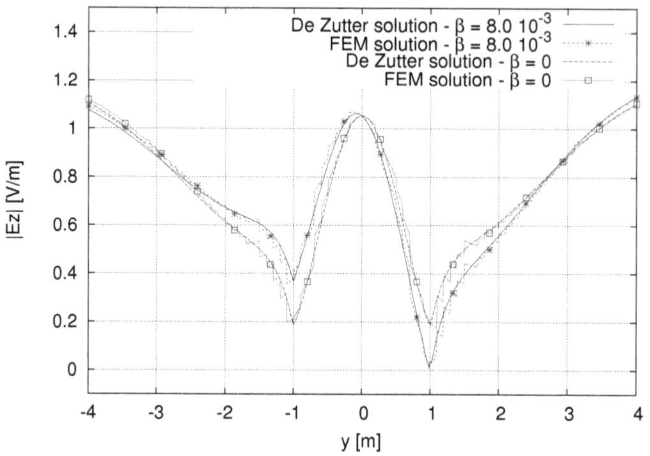

Figure 5. Comparison of the magnitudes of the z-component of the electric field along the y-axis for the rotating sphere.

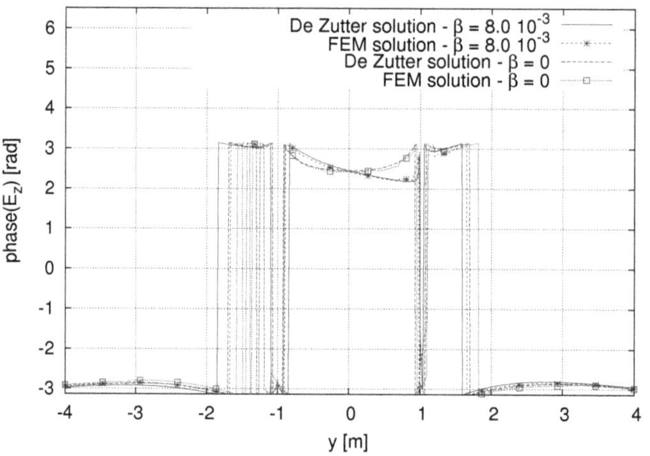

Figure 6. Comparison of the phases of the z-component of the electric field along the y-axis for the rotating sphere.

7.4. Application to Rotating Torus

Thus far, we have considered problems for which a semi-analytic solution is available. In order to illustrate the full relevance of the new results, we now tackle problems for which no solution can be found in the open literature, to the best of the authors' knowledge.

For this, let us consider a homogeneous torus rotating about its axis. The geometry of the problem is described in Figure 7.

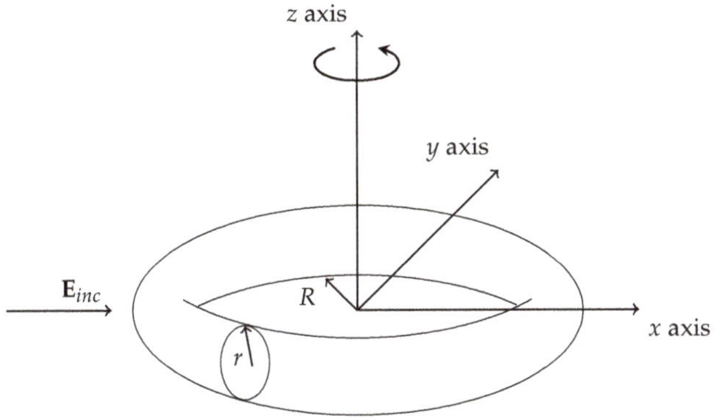

Figure 7. Geometry of the toroidal scatterer. The toroid rotates about the z-axis with angular velocity ω_s. R and r are as shown in the figure and are respectively the "major radius" and the "minor radius" of the torus.

The value of both radii (R and r) is 0.15 m. The torus is made of a material with $\varepsilon_r = 20$. The domain of numerical investigation is a sphere of radius 2 m. We consider a plane wave incident along the x-axis with the electric field polarized along the z-axis and with magnitude 1 V/m and frequency $f = 500$ MHz. For $\varepsilon_r = 20$, the upper bounds for β allowing the application of our theory are given by $\beta_{r1} = 2.618 \times 10^{-3}$ and $\beta_{critical} = 1.893 \times 10^{-3}$. Respecting these limits lets us consider values of $\omega_s \leq 4.0 \times 10^{-3} c_0$ rad/s, which corresponds to a maximum β value of 1.8×10^{-3}.

The first order edge element based Galerkin finite element solution we show in the following is obtained with a three-dimensional tetrahedral mesh having 2,192,940 elements and 36,993 nodes.

To gain an understanding of the solution, we may consider the behaviours of the field along the three coordinate directions for different rotating speed values. Here, we consider ω_s in the set $\{0, 1.0 \times 10^{-3} c_0, 2.0 \times 10^{-3} c_0, 4.0 \times 10^{-3} c_0\}$. The electric field components along the x-axis are shown in Figure 8.

In this case, the largest effect due to motion occurs in the z component of the field, where a difference as large as twice the incident field can be observed between the cases with $\omega_s = 0$ and $\omega_s = 4.0 \times 10^{-3} c_0$. For the other two speeds considered, the effects are smaller but still discernible. There are effects also on the components $|E_x|$ and $|E_y|$ along the x-axis, the maximum difference from the stationary solution being around twice the incident field in the former case and fifty percent of the incident field in the latter one. The norm of the total field $|\mathbf{E}|$ is dominated by the z-component and hence both of them carry roughly the same information when plotted along the x-axis. Along the y-axis for $\omega_s = 4.0 \times 10^{-3} c_0$, the differences from the stationary case are as large as twice the incident field for $|E_x|$, fifty percent of incident field for $|E_y|$ and three times for $|E_z|$. This is shown in Figure 9.

In this case, the total field $|\mathbf{E}|$ is also largely similar to the z component and the difference from the stationary solution is about three times the incident field. For the other speeds considered, the rotational effects on the fields are quite small along this direction. Finally, we do not show the behaviour of the electric field along the z-axis because, in this case, the effects due to motion for all the components are quite small (less than 2 percent) for the speeds considered.

Hence, we can conclude that in this case the fields along x- and y -directions carry significant information about the rotating speed of the toroidal scatterer.

As previously mentioned, the changes in the fields induced by the motion are important because it may be useful for the reconstruction of the velocity profiles of rotating objects. This could be of interest, for example, for rotating celestial bodies. Moreover, since our theory guarantees the well posedness of the problems and the convergence of the numerical solutions, the presented results can be considered as benchmarks for other approaches or numerical techniques.

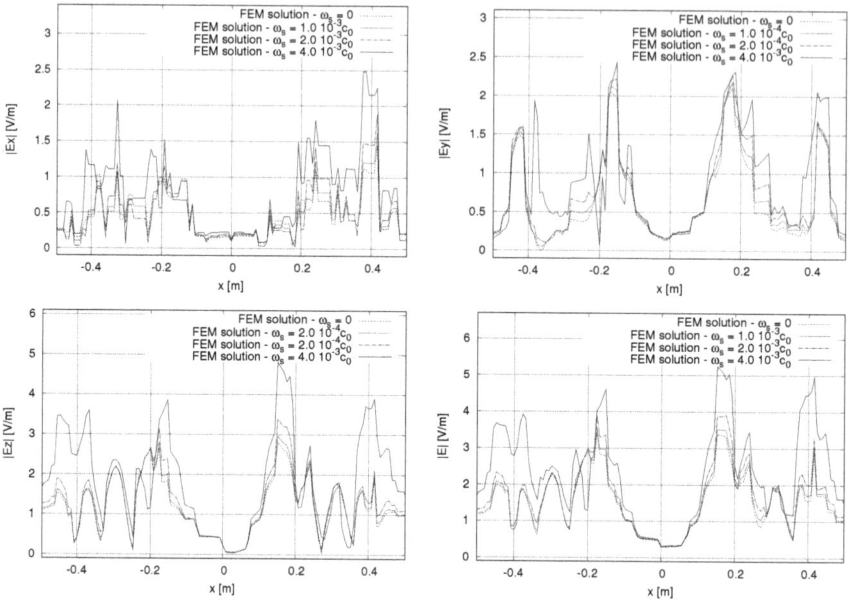

Figure 8. Magnitude of the electric field along the x-axis for different values of ω_s for rotating torus.

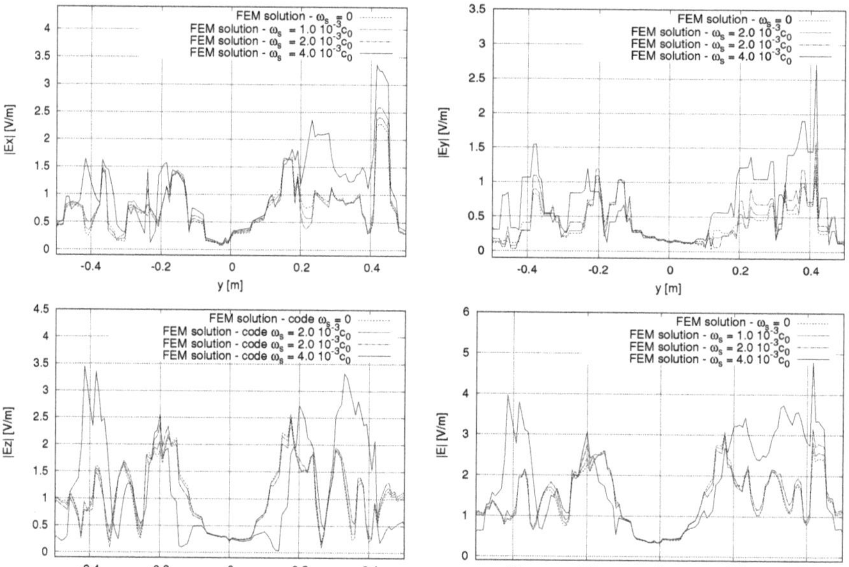

Figure 9. Magnitude of the electric field along the y-axis for different values of ω_s for rotating torus.

8. Conclusions

In this work, we have presented sufficient conditions for well posedness and finite element approximability of three-dimensional time-harmonic electromagnetic boundary value problems involving bianisotropic media. The theory is applied to electromagnetic problems involving rotating axisymmetric objects. For some of them, the solutions are not present in the open literature and, hence, they can be used as benchmarks for other approaches.

Author Contributions: Both the authors have contributed equally to this paper. All authors have read and agreed to the published version of the manuscript.

Funding: This research received no external funding.

Conflicts of Interest: The authors declare no conflict of interest.

Appendix A

In this Appendix A, we provide Lemma A1 and the proofs of Theorem 1, Theorem 2, Theorem 4, and Lemma 1.

Lemma A1. *Any solution* $\mathbf{E}$ *of Problem 2 with* $l = 0$ *satisfies*

- $\mathbf{n} \times \mathbf{E} = 0$ *on* Γ_l *if HM3, HB2 and HB3 hold true,*
- $\mathbf{E} = 0$ *in* D *if HM3, HB2 and HM4 hold true; the same result is achieved under hypotheses HM3, HB2 and HM5 or HM3, HB2, HM6 and HM7.*

Proof. Consider $\mathbf{E}$ as the solution of Problem 2 with $l = 0$ and choose $\mathbf{v} = \mathbf{E}$ in Equation (6). Since $a(\mathbf{E}, \mathbf{E}) = 0$, we get

$$0 = Im(a(\mathbf{E}, \mathbf{E})) = -\int_\Omega (\mathbf{E}^*, \text{curl}\,\mathbf{E}^*) A_{ss} \begin{pmatrix} \mathbf{E} \\ \text{curl}\,\mathbf{E} \end{pmatrix} + \omega \int_\Gamma Re(Y)|\mathbf{n} \times \mathbf{E} \times \mathbf{n}|^2. \tag{A1}$$

Taking account that $\omega > 0$, if we assume HM3, HB2, and HB3, we easily get:

$$0 \geq \int_\Gamma Re(Y)|\mathbf{n} \times \mathbf{E} \times \mathbf{n}|^2 \geq \int_{\Gamma_l} Re(Y)|\mathbf{n} \times \mathbf{E} \times \mathbf{n}|^2 \geq C_{Ym} \int_{\Gamma_l} |\mathbf{n} \times \mathbf{E} \times \mathbf{n}|^2. \tag{A2}$$

Thus, under the indicated hypotheses, we can conclude that $\mathbf{n} \times \mathbf{E} = 0$ on Γ_l.

If HM3, HB2, and HM4 hold true, considering that $\omega > 0$, we get

$$0 \geq -\int_\Omega (\mathbf{E}^*, \text{curl}\,\mathbf{E}^*) A_{ss} \begin{pmatrix} \mathbf{E} \\ \text{curl}\,\mathbf{E} \end{pmatrix} \geq K_{dl} \int_D (|\mathbf{E}|^2 + |\text{curl}\,\mathbf{E}|^2) \geq K_{dl} \int_D |\mathbf{E}|^2 \tag{A3}$$

and we conclude that $\mathbf{E} = 0$ in D.

The same result easily follows if we assume HM3, HB2 and HM5, since

$$0 \geq -\int_\Omega (\mathbf{E}^*, \text{curl}\,\mathbf{E}^*) A_{ss} \begin{pmatrix} \mathbf{E} \\ \text{curl}\,\mathbf{E} \end{pmatrix} \geq K_{el} \int_D |\mathbf{E}|^2, \tag{A4}$$

or if HM3, HB2, HM6 and HM7 hold true, since

$$0 \geq -\int_\Omega (\mathbf{E}^*, \text{curl}\,\mathbf{E}^*) A_{ss} \begin{pmatrix} \mathbf{E} \\ \text{curl}\,\mathbf{E} \end{pmatrix} \geq K_{ml} \int_D |\text{curl}\,\mathbf{E}|^2 \tag{A5}$$

and taking account of $(5)_2$ with $\mathbf{J}_m = 0$, $(1)_2$ with $M = 0$, $(5)_1$ with $\mathbf{J}_e = 0$ and $(1)_1$ with $L = 0$. □

Proof of Theorem 1. By Lemma A1, there is either a subdomain D where the electric field $\mathbf{E} = 0$ or a part of the boundary, Γ_l, where $\mathbf{n} \times \mathbf{E} = 0$. We prove that the fields $\mathbf{E}$, $\mathbf{B}$, $\mathbf{H}$ and $\mathbf{D}$ are identically zero in Ω_i, for all $i \in I$, if one of the following is true:

- Ω_i is adjacent to a region Ω_k, $k \in I$, where it has already been proved that $\mathbf{E} = 0$,
- $\partial\Omega_i$ shares a non-empty, open, Lipschitz continuous part with Γ_l.

In both cases, we introduce a sufficiently small open ball $B \subset \mathbb{R}^3$ centered on a point of $\overline{\Omega}_i \cap \overline{\Omega}_k$ or on a point of $\partial\Omega_i \cap \Gamma_l$.

In both cases, we have $\mathbf{n} \times \mathbf{E} = 0$ on $B \cap \partial\Omega_i$. Then, considering the homogeneous version of (5), by $(5)_3$ we get $\mathbf{n} \times \mathbf{H} = 0$ on $B \cap \partial\Omega_i$. Then, Equations $(5)_1$ and $(5)_2$ respectively imply that the normal components $\mathbf{n} \cdot \mathbf{D} = 0$ and $\mathbf{n} \cdot \mathbf{B} = 0$ on $B \cap \partial\Omega_i$.

Now, we can extend in an analytic way, from Ω_i to $B \setminus \Omega_i$, all the components of κ, χ, γ and ν. This is possible because of HM8. In both cases, we have to consider that $\mathbf{E}$, $\mathbf{B}$, $\mathbf{H}$ and $\mathbf{D}$ are either trivial fields in $B \setminus \Omega_i$ (in the first of the two cases) or can be trivially extended to $B \setminus \Omega_i$ (in the second of the two cases).

Next, we can show that the fields in B are analytic in Ω_i, $\forall i \in I$. As a matter of fact:

- the fields $\mathbf{E}, \mathbf{B}, \mathbf{H}, \mathbf{D}$ satisfy (5) in B since (5) holds true in Ω_i and in $B \setminus \overline{\Omega}_i$ for any $i \in I$,
- the fields $\mathbf{E}, \mathbf{B}, \mathbf{H}, \mathbf{D}$ satisfy (1) in B since (1) holds true in Ω_i and in $B \setminus \overline{\Omega}_i$ for any $i \in I$,
- by using the properties of the fields on the boundary deduced above we easily conclude that in both the cases of interest $(\mathbf{E}, \mathbf{B}, \mathbf{H}, \mathbf{D}) \in H(\mathrm{curl}, B) \times H(\mathrm{div}^0, B) \times H(\mathrm{curl}, B) \times H(\mathrm{div}^0, B)$, [24] (p. 107)
- HS1 of [22] is satisfied in any case since the sources are trivial,
- for any Ω_i, $i \in I_a$, κ and ν satisfy HM1, HM5 and HM6 of [22] in B since we have verified them for $\overline{\Omega}_i$ (by HM8, HM9 and HM10 above) and all the extended quantities are at least continuous in $\overline{\Omega}_i$ (by HM8),
- for any Ω_i, $i \in I_b$, HM1, HM8 and HM9 of [22] are satisfied in B by κ, χ, γ, ν, which are extended analytically to a sufficiently small ball B, since we have verified them in $\overline{\Omega}_i$ for all $i \in I_b$ (by HM8, HM9 and HM11 above) and all the extended quantities are at least continuous in $\overline{\Omega}_i$ (by HM8),
- for any Ω_i, $i \in I_b$, HM12 implies that (7.11) of [22] is satisfied in B by κ, χ, γ, ν extended as indicated above since we have verified it in $\overline{\Omega}_i$ for all $i \in I_b$ and all the extended quantities involved are at least continuous in $\overline{\Omega}_i$ (by HM8).

Thus, by Theorems 6.4 and 7.3 of [22], we can conclude that the electromagnetic fields in B are analytic. Since they are equal to zero or can be extended to zero fields in $B \setminus \overline{\Omega}_i$, we get $\mathbf{E} = 0$, $\mathbf{B} = 0$, $\mathbf{H} = 0$, and $\mathbf{D} = 0$ in B. Once the fields are proved to be equal to zero in B, we easily see that they are zero in Ω_i by the analyticity of the indicated fields in Ω_i. This procedure can be successively applied to all subdomains allowing us to conclude that the homogeneous version of Problem 2 has only trivial solutions and hence Problem 2 admits a unique solution. □

Proof of Theorem 2. The homogeneous variational problem defined in the statement of the theorem is similar in form to the homogeneous version of the original problem. The only differences are the reversed roles of $\mathbf{u}$ and $\mathbf{v}$ and the change in sign of the imaginary part. Hence, the same proof will also work here. In particular, in the proof of Lemma A1, we can use the fact that $(a(\mathbf{E}, \mathbf{E}))^* = 0$ implies $Im\,(a(\mathbf{E}, \mathbf{E})) = 0$, which in turn implies (A1) and hence the conclusions of Lemma A1 hold also for the homogeneous variational problem defined in the statement of the theorem. The arguments for showing the unique continuation results are not affected by the sign of imaginary part of the sesquilinear form. Hence, we can conclude that $\mathbf{v} = 0$ is the only solution. □

Proof of Theorem 4. We prove Theorem 4 by contradiction, as we did in [10]. Due to the similarities with the corresponding proof presented in [10], we report here the main ideas.

As in [10], we get the result by contradiction and, thus, we assume that:

$$\exists \{\mathbf{u}_n\},\ \mathbf{u}_n \in U \text{ and } \|\mathbf{u}_n\|_U = 1\ \forall n \in \mathbb{N}, \text{ such that } \lim_{n\to\infty} \sup_{\|\mathbf{v}\|_U \le 1} |a(\mathbf{u}_n, \mathbf{v})| = 0. \tag{A6}$$

For the space U, under hypotheses HD1–HD3, HM1–HM2, and HM13, the following Helmholtz decomposition holds true [24] (p. 86):

$$U = U_0 \oplus U_1, \tag{A7}$$

where

$$U_0 = \{\mathbf{u} \in U \mid \operatorname{curl} \mathbf{u} = 0 \text{ in } \Omega \text{ and } \mathbf{u} \times \mathbf{n} = 0 \text{ on } \Gamma\} \tag{A8}$$

and

$$U_1 = \{\mathbf{u} \in U \mid (P\mathbf{u}, \mathbf{v})_{0,\Omega} = 0 \; \forall \mathbf{v} \in U_0\}. \tag{A9}$$

Thus, for any element of the sequence satisfying (A6), we get

$$\mathbf{u}_n = \mathbf{u}_{n0} + \mathbf{u}_{n1}, \tag{A10}$$

with $\mathbf{u}_{n0} \in U_0$ and $\mathbf{u}_{n1} \in U_1$.

Under the assumed hypotheses, one easily gets:

$$\|\mathbf{u}_{n0}\|_{0,\Omega} = \|\mathbf{u}_{n0}\|_U \le \frac{C_{PL}}{C_{PS}} \|\mathbf{u}_n\|_{0,\Omega} \le \frac{C_{PL}}{C_{PS}} \|\mathbf{u}_n\|_U, \tag{A11}$$

$$\|\mathbf{u}_{n1}\|_U \le \frac{C_{PS} + C_{PL}}{C_{PS}} \|\mathbf{u}_n\|_U, \tag{A12}$$

$$\lim_{n\to\infty} \|\mathbf{n} \times \mathbf{u}_n \times \mathbf{n}\|_{0,\Gamma} = 0. \tag{A13}$$

Thus, the two sequences, $\{\mathbf{u}_{n0}\}$ and $\{\mathbf{u}_{n1}\}$, subsequences of the sequence satisfying (A6), are bounded in U. Since our hypotheses guarantees that uniqueness also holds, then, on a common subsequence of indices, both $\{\mathbf{u}_{n0}\}$ and $\{\mathbf{u}_{n1}\}$ weakly converge to zero in U and, by the compact embedding of U_1 in $(L^2(\Omega))^3$, which holds true under hypothesis HM13, from $\{\mathbf{u}_{n1}\}$, we can extract a subsequence which converges strongly in $(L^2(\Omega))^3$ to $\hat{\mathbf{u}}_1$. Finally, since both weak convergence in U and strong convergence in $(L^2(\Omega))^3$ imply weak convergence in $(L^2(\Omega))^3$ to the same limit, we immediately deduce $\hat{\mathbf{u}}_1 = 0$.

By setting $\mathbf{u} = \mathbf{u}_n$ and $\mathbf{v} = \mathbf{u}_{n0}$ for any n, we get from the very definition of the sesquilinear form a:

$$C_{PS}\|\mathbf{u}_{n0}\|^2_{0,\Omega} \le \frac{c_0}{\omega^2}|a(\mathbf{u}_n, \mathbf{u}_{n0})| + \frac{c_0 C_L}{\omega}\|\operatorname{curl} \mathbf{u}_n\|_{0,\Omega}\|\mathbf{u}_{n0}\|_{0,\Omega}. \tag{A14}$$

By the same token, by setting $\mathbf{u} = \mathbf{u}_n$ and $\mathbf{v} = \mathbf{u}_{n1}$ for any n, we deduce

$$\begin{aligned} &c_0 C_{QS}\|\operatorname{curl} \mathbf{u}_{n1}\|^2_{0,\Omega} \le \\ &|a(\mathbf{u}_n, \mathbf{u}_{n1})| + \frac{\omega^2 C_{PL}}{c_0}\|\mathbf{u}_n\|_{0,\Omega}\|\mathbf{u}_{n1}\|_{0,\Omega} + \omega C_M \|\mathbf{u}_{n0}\|_{0,\Omega}\|\operatorname{curl} \mathbf{u}_{n1}\|_{0,\Omega} + \\ &+\omega(C_M + C_L)\|\mathbf{u}_{n1}\|_{0,\Omega}\|\operatorname{curl} \mathbf{u}_{n1}\|_{0,\Omega} + \omega C_{YL}\|\mathbf{n} \times \mathbf{u}_{n1} \times \mathbf{n}\|^2_{0,\Gamma}. \end{aligned} \tag{A15}$$

Now, taking into account that $\{\mathbf{u}_{n0}\}$ and $\{\mathbf{u}_{n1}\}$ are bounded in U, $\|\mathbf{u}_{n1}\|_{0,\Omega} \to 0$ on a subsequence, $\|(\mathbf{n} \times \mathbf{u}_{n1} \times \mathbf{n})\|_{0,\Gamma} \to 0$, by using inequalities (A14) and (A15), we deduce that we cannot find a subsequence such that either $\{\mathbf{u}_{n0}\}$ or $\{\operatorname{curl} \mathbf{u}_{n1}\}$ converges to zero in $(L^2(\Omega))^3$. As a matter of fact, if one of them did converge to zero in $(L^2(\Omega))^3$, then both should do and we would obtain that $\{\mathbf{u}_n\}$ should converge to zero in U against the hypothesis.

Then, we can find a subsequence giving $\|\mathbf{u}_{n1}\|_{0,\Omega} \to 0$ and $\|\mathbf{u}_{n0}\|_{0,\Omega} \ge \epsilon > 0$. On this subsequence, from inequality (A14), we get

$$\|\mathbf{u}_{n0}\|_{0,\Omega} \le \frac{c_0}{\omega^2 C_{PS}}|a(\mathbf{u}_n, \frac{\mathbf{u}_{n0}}{\|\mathbf{u}_{n0}\|_{0,\Omega}})| + \frac{c_0 C_L}{\omega C_{PS}}\|\operatorname{curl} \mathbf{u}_{n1}\|_{0,\Omega}. \tag{A16}$$

By substituting the right-hand side of (A16) for $\|\mathbf{u}_{n0}\|_{0,\Omega}$ in inequality (A15), we deduce

$$
\begin{aligned}
&c_0\left(C_{QS}-\frac{C_LC_M}{C_{PS}}\right)\|\mathrm{curl}\,\mathbf{u}_{n1}\|_{0,\Omega}^2 \leq \\
&\frac{C_{PS}+C_{PL}}{C_{PS}}\left|a\left(\mathbf{u}_n,\frac{C_{PS}}{C_{PS}+C_{PL}}\mathbf{u}_{n1}\right)\right| + \frac{\omega^2 C_{PL}}{c_0}\|\mathbf{u}_n\|_{0,\Omega}\|\mathbf{u}_{n1}\|_{0,\Omega} + \\
&+\omega(C_M+C_L)\|\mathbf{u}_{n1}\|_{0,\Omega}\|\mathrm{curl}\,\mathbf{u}_{n1}\|_{0,\Omega}+\omega C_{YL}\|(\mathbf{n}\times\mathbf{u}_{n1}\times\mathbf{n})\|_{0,\Gamma}^2 + \\
&+\frac{c_0C_M}{\omega C_{PS}}\left|a\left(\mathbf{u}_n,\frac{\mathbf{u}_{n0}}{\|\mathbf{u}_{n0}\|_{0,\Omega}}\right)\right|\|\mathrm{curl}\,\mathbf{u}_{n1}\|_{0,\Omega}.
\end{aligned} \tag{A17}
$$

The right-hand side of inequality (A17) converges to zero on the indicated subsequence and, by hypothesis HM15, we get $\|\mathrm{curl}\,\mathbf{u}_{n1}\|_{0,\Omega}\to 0$, which is against the starting hypothesis. □

Proof of Lemma 1. We have to analyse just the case when Ω_{el} is neither the whole Ω nor the empty set. For all $\mathbf{u}\in(L^2(\Omega))^3$, we have

$$
\begin{aligned}
|(P\mathbf{u},\mathbf{u})_{0,\Omega}|^2 &= \left|\int_\Omega \mathbf{u}^*P_s\mathbf{u} - j\int_\Omega \mathbf{u}^*P_{ss}\mathbf{u}\right|^2 = \\
&= \left(\int_\Omega \mathbf{u}^*P_s\mathbf{u}\right)^2 + \left(\int_\Omega \mathbf{u}^*P_{ss}\mathbf{u}\right)^2 = \\
&= \left(\int_{\Omega\setminus\Omega_{el}} \mathbf{u}^*P_s\mathbf{u} - \int_{\Omega_{el}} -\mathbf{u}^*P_s\mathbf{u}\right)^2 + \left(\int_{\Omega_{el}} \mathbf{u}^*P_{ss}\mathbf{u} + \int_{\Omega\setminus\Omega_{el}} \mathbf{u}^*P_{ss}\mathbf{u}\right)^2.
\end{aligned} \tag{A18}
$$

Under assumption HM3, by using Lemma B.1 of [9] with $K_1=K_2=0$, we get that P_{ss} is positive semi definite in Ω_i, $\forall i\in I$. Moreover, since Ω_{el} is the union of the subdomains Ω_i of Ω where P_{ss} is uniformly positive definite, we get

$$
|(P\mathbf{u},\mathbf{u})_{0,\Omega}|^2 \geq \left(\int_{\Omega\setminus\Omega_{el}} \mathbf{u}^*P_s\mathbf{u} - \int_{\Omega_{el}} -\mathbf{u}^*P_s\mathbf{u}\right)^2 + C_1^2||\mathbf{u}||_{0,\Omega_{el}}^4. \tag{A19}
$$

However, for all $a,b\in\mathbb{R}$, for any $\alpha>0$, we have

$$
(a-b)^2 \geq (1-\alpha)a^2 + (1-\frac{1}{\alpha})b^2. \tag{A20}
$$

Then, using the above inequality for the first addend of the right-hand side of Equation (A19), we get

$$
|(P\mathbf{u},\mathbf{u})_{0,\Omega}|^2 \geq (1-\alpha)\left(\int_{\Omega\setminus\Omega_{el}} \mathbf{u}^*P_s\mathbf{u}\right)^2 + (1-\frac{1}{\alpha})\left(\int_{\Omega_{el}} \mathbf{u}^*P_s\mathbf{u}\right)^2 + C_1^2||\mathbf{u}||_{0,\Omega_{el}}^4. \tag{A21}
$$

The validity of assumption HM2 guarantees that inequality (28) holds true. Then, by taking account that $1-\frac{1}{\alpha}<0$ for all $\alpha\in(0,1)$, we get

$$
|(P\mathbf{u},\mathbf{u})_{0,\Omega}| \geq (1-\alpha)\left(\int_{\Omega\setminus\Omega_{el}} \mathbf{u}^*P_s\mathbf{u}\right)^2 + (C_1^2+(1-\frac{1}{\alpha})C_3^2)||\mathbf{u}||_{0,\Omega_{el}}^4. \tag{A22}
$$

By using (25), we then deduce

$$
|(P\mathbf{u},\mathbf{u})_{0,\Omega}|^2 \geq (1-\alpha)C_5^2||\mathbf{u}||_{0,\Omega\setminus\Omega_{el}}^4 + (C_1^2+(1-\frac{1}{\alpha})C_3^2)||\mathbf{u}||_{0,\Omega_{el}}^4. \tag{A23}
$$

By defining $1 > \alpha > \frac{C_3^2}{C_1^2+C_3^2} > 0$, we have that both terms in (A23) are positive. As a matter of fact, we can think of the right-hand side of (A23) as $s^2 + t^2$, $s, t \in \mathbb{R}$, and, since $s^2 + t^2 \geq \frac{(s+t)^2}{2}$, we get

$$|(P\mathbf{u},\mathbf{u})_{0,\Omega}|^2 \geq \frac{1}{2}\left(\sqrt{(1-\alpha)}\,C_5||\mathbf{u}||^2_{0,\Omega\setminus\Omega_{el}} + \sqrt{C_1^2 + (1-\frac{1}{\alpha})C_3^2}\,||\mathbf{u}||^2_{0,\Omega_{el}}\right)^2 \geq$$
$$\geq \frac{1}{2}\left(\min\left(\sqrt{(1-\alpha)}\,C_5, \sqrt{C_1^2 + (1-\frac{1}{\alpha})C_3^2}\right)\right)^2 \left(||\mathbf{u}||^2_{0,\Omega\setminus\Omega_{el}} + ||\mathbf{u}||^2_{0,\Omega_{el}}\right)^2 = \tag{A24}$$
$$= \frac{1}{2}\min\left((1-\alpha)C_5^2, C_1^2 + (1-\frac{1}{\alpha})C_3^2\right)||\mathbf{u}||^4_{0,\Omega}.$$

□

References

1. De Zutter, D. Scattering by a rotating dielectric sphere. *IEEE Trans. Antennas Propag.* **1980**, *28*, 643–651. [CrossRef]
2. Van Bladel, J. Rotating dielectric sphere in a low-frequency field. *Proc. IEEE* **1979**, *67*, 1654–1655. [CrossRef]
3. Cheng, D.K.; Kong, J.A. Covariant descriptions of bianisotropic media. *Proc. IEEE* **1968**, *56*, 248–251. [CrossRef]
4. Sommerfeld, A. *Electrodynamics*; Lectures on Theoretical Physics; Academic Press: New York, NY, USA, 1952.
5. Kraft, M.; Braun, A.; Luo, Y.; Maier, S.A.; Pendry, J.B. Bianisotropy and magnetism in plasmonic gratings. *ACS Photonics* **2016**, *3*, 764–769. [CrossRef]
6. Yazdi, M.; Albooyeh, M.; Alaee, R.; Asadchy, V.; Komjani, N.; Rockstuhl, C.; Simovski, C.R.; Tretyakov, S. A bianisotropic metasurface with resonant asymmetric absorption. *IEEE Trans. Antennas Propag.* **2015**, *63*, 3004–3015. [CrossRef]
7. Kildishev, A.V.; Borneman, J.D.; Ni, X.; Shalaev, V.M.; Drachev, V.P. Bianisotropic effective parameters of optical metamagnetics and negative-index materials. *Proc. IEEE* **2011**, *99*, 1691–1700. [CrossRef]
8. Kriegler, C.E.; Rill, M.S.; Linden, S.; Wegener, M. Bianisotropic photonic metamaterials. *IEEE J. Sel. Top. Quantum Electron.* **2010**, *16*, 367–375. [CrossRef]
9. Fernandes, P.; Raffetto, M. Well-posedness and finite element approximability of time-harmonic electromagnetic boundary value problems involving bianisotropic materials and metamaterials. *Math. Model. Methods Appl. Sci.* **2009**, *19*, 2299–2335. [CrossRef]
10. Brignone, M.; Raffetto, M. Well posedness and finite element approximability of two-dimensional time-harmonic electromagnetic problems involving non-conducting moving objects with stationary boundaries. *ESAIM Math. Model. Numer. Anal.* **2015**, *49*, 1157–1192. [CrossRef]
11. Ioannidis, A.D.; Kristensson, G.; Stratis, I.G. On the well-posedness of the Maxwell system for linear bianisotropic media. *SIAM J. Math. Anal.* **2012**, *44*, 2459–2473. [CrossRef]
12. Cocquet, P.; Mazet, P.; Mouysset, V. On the existence and uniqueness of a solution for some frequency-dependent partial differential equations coming from the modeling of metamaterials. *SIAM J. Math. Anal.* **2012**, *44*, 3806–3833. [CrossRef]
13. Costen, R.C.; Adamson, D. Three-dimensional derivation of the electrodynamic jump conditions and momentum-energy laws at a moving boundary. *Proc. IEEE* **1965**, *53*, 1181–1196. [CrossRef]
14. Bilotti, F.; Vegni, L.; Toscano, A. Radiation and scattering features of patch antennas with bianisotropic substrates. *IEEE Trans. Antennas Propag.* **2003**, *51*, 449–456. [CrossRef]
15. Cheng, X.; Chen, H.; Wu, B.I.; Kong, J.A. Cloak for Bianisotropic and Moving Media. *Prog. Electromagn. Res.* **2009**, *89*, 199–212. [CrossRef]
16. Alotto, P.; Codecasa, L. A fit formulation of bianisotropic materials over polyhedral grids. *IEEE Trans. Magn.* **2014**, *50*, 349–352. [CrossRef]
17. Wu, T.X.; Jaggard, D.L. A comprehensive study of discontinuities in chirowaveguides. *IEEE Trans. Microw. Theory Tech.* **2002**, *50*, 2320–2330. [CrossRef]

18. Kalarickel Ramakrishnan, P.; Raffetto, M. Accuracy of finite element approximations for two-dimensional time-harmonic electromagnetic boundary value problems involving non-conducting moving objects with stationary boundaries. *ACES J.* **2018**, *33*, 585–596.
19. Van Bladel, J.G. *Electromagnetic Fields*, 2nd ed.; IEEE Press: Piscataway, NJ, USA, 2007.
20. Harrington, R.F. *Time-Harmonic Electromagnetic Fields*; McGraw-Hill: New York, NY, USA, 1961.
21. Kong, J.A. *Theory of Electromagnetic Waves*; Wiley: New York, NY, USA, 1975.
22. Fernandes, P.; Ottonello, M.; Raffetto, M. Regularity of time-harmonic electromagnetic fields in the interior of bianisotropic materials and metamaterials. *IMA J. Appl. Math.* **2014**, *79*, 54–93. [CrossRef]
23. Girault, V.; Raviart, P.A. *Finite Element Methods for Navier–Stokes Equations*; Springer–Verlag: Berlin, Germany, 1986.
24. Monk, P. *Finite Element Methods for Maxwell's Equations*; Oxford Science Publications: Oxford, UK, 2003.
25. Taylor, A.E. *Introduction to Functional Analysis*; John Wiley & Sons: New York, NY, USA, 1958.
26. Leis, R., *Trends in Applications of Pure Mathematics to Mechanics*; Chapter Exterior Boundary-Value Problems In Mathematical Physics; Pitman: London, UK, 1979; Volume 11, pp. 187–203.
27. Hazard, C.; Lenoir, M. On the solution of time-harmonic scattering problems for Maxwell's equations. *SIAM J. Math. Anal.* **1996**, *27*, 1597–1630. [CrossRef]
28. Alonso, A.; Raffetto, M. Unique solvability for electromagnetic boundary value problems in the presence of partly lossy inhomogeneous anisotropic media and mixed boundary conditions. *Math. Model. Methods Appl. Sci.* **2003**, *13*, 597–611. [CrossRef]
29. Ciarlet, P.G. *The Finite Element Method for Elliptic Problems*; North-Holland: Amsterdam, The Netherlands, 1978.
30. Caorsi, S.; Fernandes, P.; Raffetto, M. On the convergence of Galerkin finite element approximations of electromagnetic eigenproblems. *SIAM J. Numer. Anal.* **2000**, *38*, 580–607. [CrossRef]
31. Caorsi, S.; Fernandes, P.; Raffetto, M. Spurious-free approximations of electromagnetic eigenproblems by means of Nedelec-type elements. *Math. Model. Numer. Anal.* **2001**, *35*, 331–354. [CrossRef]
32. Jin, J. *The Finite Element Method in Electromagnetics*; John Wiley & Sons: New York, NY, USA, 1993.
33. Barrett, R.; Berry, M.; Chan, T.F.; Demmel, J.; Donato, J.; Dongarra, J.; Eijkhout, V.; Pozo, R.; Romine, C.; der Vorst, H.V. *Templates for the Solution of Linear Systems: Building Blocks for Iterative Methods*, 2nd ed.; SIAM: Philadelphia, PA, USA, 1994.
34. Fernandes, P.; Raffetto, M. Existence, uniqueness and finite element approximation of the solution of time-harmonic electromagnetic boundary value problems involving metamaterials. *COMPEL* **2005**, *24*, 1450–1469. [CrossRef]
35. De Zutter, D. Scattering by a rotating circular cylinder with finite conductivity. *IEEE Trans. Antennas Propag.* **1983**, *31*, 166–169. [CrossRef]
36. Brignone, M.; Ramakrishnan, P.K.; Raffetto, M. A first numerical assessment of the reliability of finite element simulators for time-harmonic electromagnetic problems involving rotating axisymmetric objects. In Proceedings of the 2016 URSI International Symposium on Electromagnetic Theory (EMTS), Espoo, Finland, 14 August 2016; pp. 787–790.
37. Pastorino, M.; Raffetto, M.; Randazzo, A. Electromagnetic inverse scattering of axially moving cylindrical targets. *IEEE Trans. Geosci. Remote Sens.* **2015**, *53*, 1452–1462. [CrossRef]
38. Brignone, M.; Gragnani, G.L.; Pastorino, M.; Raffetto, M.; Randazzo, A. Noise limitations on the recovery of average values of velocity profiles in pipelines by simple imaging systems. *IEEE Geosci. Remote Sens. Lett.* **2016**, *13*, 1340–1344. [CrossRef]

symmetry

MDPI

Article

Free Vibration Analysis of Functionally Graded Shells Using an Edge-Based Smoothed Finite Element Method

Tien Dat Pham [1], Quoc Hoa Pham [1], Van Duc Phan [2], Hoang Nam Nguyen [3,*] and Van Thom Do [1]

1 Faculty of Mechanical Engineering, Le Quy Don Technical University, Ha Noi 100000, Vietnam; tiendat1962@gmail.com (T.D.P.); quochoavihempich@gmail.com (Q.H.P.); thom.dovan.mta@gmail.com (V.T.D.)
2 Center of Excellence for Automation and Precision Mechanical Engineering, Nguyen Tat Thanh University, Ho Chi Minh 700000, Vietnam; pvduc@ntt.edu.vn
3 Modeling Evolutionary Algorithms Simulation and Artificial Intelligence, Faculty of Electrical & Electronics Engineering, Ton Duc Thang University, Ho Chi Minh 700000, Vietnam
* Correspondence: nguyenhoangnam@tdtu.edu.vn

Received: 7 March 2019; Accepted: 7 May 2019; Published: 17 May 2019

Abstract: An edge-based smoothed finite element method (ES-FEM) combined with the mixed interpolation of tensorial components technique for triangular shell element (MITC3), called ES-MITC3, for free vibration analysis of functionally graded shells is investigated in this work. In the formulation of the ES-MITC3, the stiffness matrices are obtained by using the strain-smoothing technique over the smoothing domains that are formed by two adjacent MITC3 triangular shell elements sharing an edge. The strain-smoothing technique can improve significantly the accuracy and convergence of the original MITC3. The material properties of functionally graded shells are assumed to vary through the thickness direction by a power–rule distribution of volume fractions of the constituents. The numerical examples demonstrated that the present ES-MITC3method is free of shear locking and achieves the high accuracy compared to the reference solutions in the literature.

Keywords: FGMshells; edge-based smoothed finite element method (ES-FEM); mixed interpolation of tensorial components (MITC)

1. Introduction

Functionally graded materials (FGM) are usually made from a mixture of metals and ceramics, whose material properties vary smoothly and continuously from one surface to the other of the structure according to volume fraction power–law distribution. It is well known that the ceramics are capable of resisting high temperature, while the metals provide structural strength and fracture toughness. They are therefore suitable to apply for aerospace structures, nuclear plants, and other applications. With the advantageous features of the FGM in many practical applications, the problem of static and free vibration behaviors of FGM shell structures are attractive to many researchers over the world. Woo and Meguid [1] used an analytical solution based on the von Karman theory to investigate nonlinear respond of FGM plates and shallow shells. Matsunaga [2] carried out the power series expansion of displacement component approach, which relied on higher-order shear deformation theory (HSDT) to analyze free vibration and buckling of FGM shells. Nguyen et al. [3] proposed an analytical solution using Reddy's HSDT to solve nonlinear dynamic and free vibration of piezoelectric FGM double curved shallow shells subjected to electrical, thermal, mechanical, and damping loads. Dao et al. [4] presented nonlinear vibration of stiffened functionally graded double curved shells on an elastic foundation using the first order shear deformation theory (FSDT) and stress function. Due to the complication of mathematics, it is generally difficult to use analytical methods for all problems. Thus, numerical methods have been devised to study the behavior of FGM structural components. Among these numerical approaches, the finite element method has become the most powerful, reliable, and simply tool to solve FGM shells. Arciniega and Reddy [5] presented a finite element formulation for nonlinear analysis of FGM shell based on the FSDT, consisting of seven parameters. Pradyumna and Bandyopadhyay [6] used the shell element, consisting of nine degrees of freedom per node, to investigate free vibration of FGM shells. Kordkheili and Naghdabadi [7] proposed a finite element model for geometrically nonlinear thermos-elastic analysis of FGM plates and shells.

In addition, in the recent trend of development of numerical methods, the edge-based smoothed finite element method (ES-FEM) combined with the mixed interpolation of tensorial components using triangular element (MITC3), named ES-MITC3, has been proposed to investigate plate and shell structures. For instance, Chau-Dinh et al. [8] proposed an ES-MITC3 to analyze static and free vibration of plates. Nguyen et al. [9] developed the ES-MITC3 for static and vibration analysis of isotropic and functionally graded plates. Pham et al. [10] examined the static and free vibration of composite shells using ES-MITC3 shell element. Pham et al. [11] used ES-MITC3 shell element to study geometrically nonlinear analysis of FGM shells based on FSDT. Pham-Tien et al. [12] investigated the dynamic response of composite shells based on the FSDT and ES-MITC3 element. Hoang-Nam Nguyen et al. [13] used FSDT to investigate dynamic composite shell with shear connectors. For shell structures, especially the shell with two curvatures, the employing of quadrilateral elements will not accurately describe the model due to the distorting during the bending process. In this case, the using of triangle elements is suitable because they can rotate freely around their three edges. However, the using of these elements can meet the shear locking phenomenon; thus, we propose the new method, in which the triangle element is combined with an edge-based smoothed finite element method (ES-MITC3) to analyze the shell structures. Its accuracy in comparison with other methods is shown in the numerical exploration.

This paper now further extends the ES-MITC3 method for free vibration analysis of functionally graded shell structures. The material properties of functionally graded shells are assumed to vary continuously and smoothly through the thickness based on a simple power–law of the volume fractions exponents. The formulation is based on the FSDT and flat shell theory due to the simplicity and computational efficiency. The accuracy and reliability of the present method are verified by comparing with those of others available numerical results.

2. Theoretical Formulation

2.1. *Functionally Grade Material*

FGM is formed by a mixture of ceramic and metal, as shown in Figure 1. The material properties change continuously from a surface to the other surface according to a power–law of volume fraction

$$P(z) = (P_c - P_m)V_c + P_m \tag{1}$$

$$V_c(z) = \left(\frac{1}{2} + \frac{z}{h}\right)^n \tag{2}$$

where $P(z)$ represents the effective material properties: Young's modulus E, density ρ and Poisson ratio v; P_c and P_m denote the properties of the ceramic and metal, respectively; V_c is the volumefractions of the ceramic; h the thickness of structure; $n \geq 0$ the volumefraction exponent; $z \in [-h/2, h/2]$ is the thickness coordinate of the structure. Figure 2 illustrates the variation of the volume fraction of ceramic and metal through the thickness via the volumefraction exponent n.

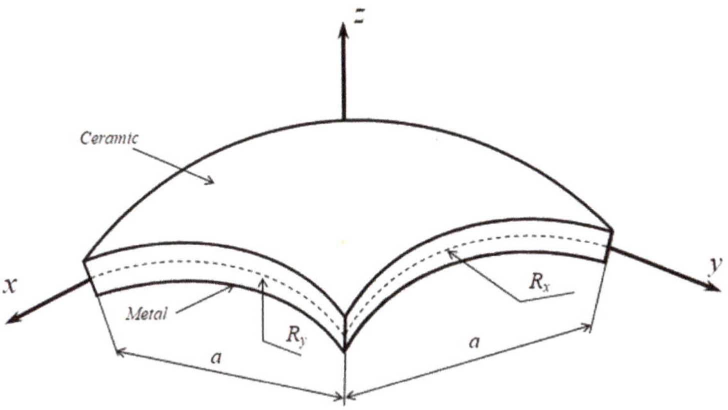

Figure 1. A functionally graded shell.

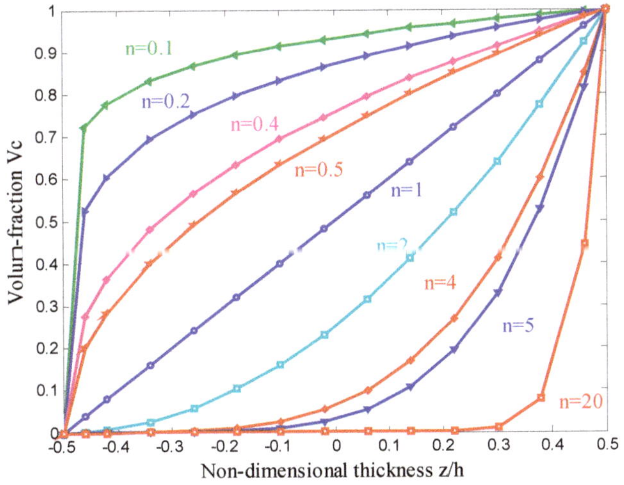

Figure 2. Variation of the volume fraction versus the non-dimensional thickness.

2.2. The FGM Shell Model

Consider an FGM shell element subjected to both in-plane forces and bending moments as shown in Figure 3. The middle (neutral) surface of the shell is chosen as the reference plane that occupies a domain $\Omega \in \Re^3$. Let u_0, v_0, and w_0 be the displacements of the middle plane in the x, y, and z directions; β_x, β_y, and β_z be the rotations of the middle surface of the shell around the y-axis, x-axis, and z-axis, respectively, as indicated in Figure 3. The unknown vector of an FGM shell including six independent variables at any point in the problem domain can be written as

$$\mathbf{u} = \begin{bmatrix} u_0 & v_0 & w_0 & \beta_x & \beta_y & \beta_z \end{bmatrix} \tag{3}$$

The linear strain–displacement relationship can be given as

$$\boldsymbol{\varepsilon} = \begin{Bmatrix} \varepsilon_x \\ \varepsilon_y \\ \gamma_{xy} \end{Bmatrix} = \boldsymbol{\varepsilon}_m + z\boldsymbol{\kappa} \tag{4}$$

$$\boldsymbol{\varepsilon}_m = \left\{ \begin{array}{c} \frac{\partial u_0}{\partial x} \\ \frac{\partial v_0}{\partial y} \\ \frac{\partial u_0}{\partial y} + \frac{\partial v_0}{\partial x} \end{array} \right\}, \boldsymbol{\kappa} = \left\{ \begin{array}{c} \frac{\partial \beta_x}{\partial x} \\ \frac{\partial \beta_y}{\partial y} \\ \frac{\partial \beta_x}{\partial y} + \frac{\partial \beta_y}{\partial x} \end{array} \right\}; \tag{5}$$

$$\boldsymbol{\gamma} = \left\{ \begin{array}{c} \gamma_{xz} \\ \gamma_{yz} \end{array} \right\} = \left\{ \begin{array}{c} \frac{\partial w_0}{\partial x} + \beta_x \\ \frac{\partial w_0}{\partial y} + \beta_y \end{array} \right\} \tag{6}$$

From Hooke's law, the constitutive relations of FGM shells are expressed as:

$$\left\{ \begin{array}{c} \sigma_{xx} \\ \sigma_{yy} \\ \tau_{xy} \\ \tau_{xz} \\ \tau_{yz} \end{array} \right\} = \left[\begin{array}{ccccc} Q_{11} & Q_{12} & 0 & 0 & 0 \\ Q_{21} & Q_{22} & 0 & 0 & 0 \\ 0 & 0 & Q_{66} & 0 & 0 \\ 0 & 0 & 0 & Q_{55} & 0 \\ 0 & 0 & 0 & 0 & Q_{44} \end{array} \right] \left\{ \begin{array}{c} \varepsilon_{xx} \\ \varepsilon_{yy} \\ \gamma_{xy} \\ \gamma_{xz} \\ \gamma_{yz} \end{array} \right\} \tag{7}$$

where

$$Q_{11}(z) = \frac{E(z)}{1-\nu(z)^2}, \; Q_{12}(z) = \nu(z)Q_{11}(z), \; Q_{22}(z) = Q_{11}(z), \; Q_{44}(z) = Q_{55}(z) = Q_{66}(z) = \frac{E(z)}{2(1+\nu(z))} \tag{8}$$

A weak form of the free vibration analysis for FGM shells can be briefly given as:

$$\int_\Omega \delta\bar{\boldsymbol{\varepsilon}}^T \mathbf{D} \bar{\boldsymbol{\varepsilon}} \mathrm{d}\Omega + \int_\Omega \delta\boldsymbol{\gamma}^T \mathbf{C} \boldsymbol{\gamma} \mathrm{d}\Omega = \int_\Omega \delta\mathbf{u}^T \mathbf{m} \ddot{\mathbf{u}} \mathrm{d}\Omega \tag{9}$$

where

$$\bar{\boldsymbol{\varepsilon}} = \left[\begin{array}{c} \boldsymbol{\varepsilon}_m \\ \boldsymbol{\kappa} \end{array} \right], \; \mathbf{D} = \left[\begin{array}{cc} \mathbf{D}^m & \mathbf{B} \\ \mathbf{B} & \mathbf{D}^b \end{array} \right] \tag{10}$$

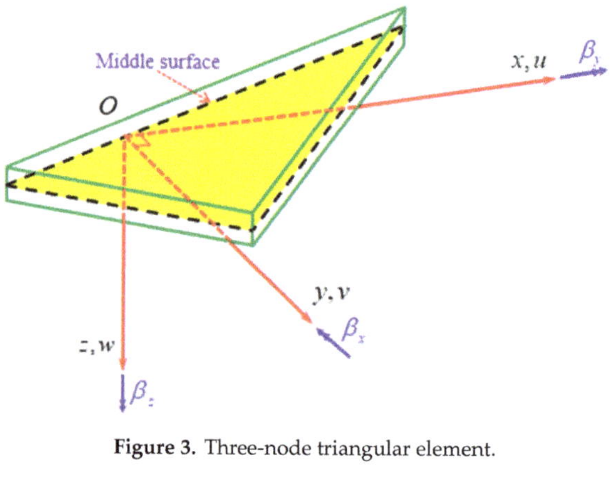

Figure 3. Three-node triangular element.

In which

$$\mathbf{D}^m = \left[\begin{array}{ccc} A_{11} & A_{12} & A_{16} \\ A_{12} & A_{22} & A_{26} \\ A_{16} & A_{26} & A_{66} \end{array} \right], \; \mathbf{B} = \left[\begin{array}{ccc} B_{11} & B_{12} & B_{16} \\ B_{12} & B_{22} & B_{26} \\ B_{16} & B_{26} & B_{66} \end{array} \right], \; \mathbf{D}^b = \left[\begin{array}{ccc} D_{11} & D_{12} & D_{16} \\ D_{12} & D_{22} & D_{26} \\ D_{16} & D_{26} & D_{66} \end{array} \right], \; \mathbf{C} = \left[\begin{array}{cc} C_{55} & C_{45} \\ C_{45} & C_{44} \end{array} \right] \tag{11}$$

In Equation (11) A_{ij}, B_{ij}, D_{ij}, and C_{ij} are given by:

$$\left(A_{ij}, B_{ij}, D_{ij}\right) = \int_{-h/2}^{h/2} Q_{ij}\left(1, z, z^2\right) dz,\ i, j = 1, 2, 6 \tag{12}$$

$$C_{ij} = k \int_{-h/2}^{h/2} Q_{ij} dz,\ i, j = 4, 5 \tag{13}$$

where $k = 5/6$ is transverse shear correction coefficient and $\mathbf{m}$ is the mass matrix containing ρ calculated as

$$\mathbf{m} = \begin{bmatrix} I_0 & 0 & 0 & I_1 & 0 & 0 \\ 0 & I_0 & 0 & 0 & I_1 & 0 \\ 0 & 0 & I_0 & 0 & 0 & 0 \\ I_1 & 0 & 0 & I_2 & 0 & 0 \\ 0 & I_1 & 0 & 0 & I_2 & 0 \\ 0 & 0 & 0 & 0 & 0 & 0 \end{bmatrix}; \ (I_0, I_1, I_2) = \int_{-h/2}^{h/2} \rho(1, z, z^2) \mathrm{d}z \tag{14}$$

2.3. Finite Element Formulation for Shell Analysis

Now, we discretize the bounded domain Ω into n^e finite three-node triangular elements with n^n nodes such that $\Omega \approx \sum_{e=1}^{n^e} \Omega_e$ and $\Omega_i \cap \Omega_j = \varnothing,\ i \neq j$. The displacement field $\mathbf{u}^e = \left\{ u^e \quad v^e \quad w^e \quad \beta_x^e \quad \beta_y^e \quad \beta_z^e \right\}^T$ of the finite element solution can be expressed as

$$\mathbf{u}^e = \sum_{i=1}^{n^n} \begin{bmatrix} N_i(x) & 0 & 0 & 0 & 0 & 0 \\ 0 & N_i(x) & 0 & 0 & 0 & 0 \\ 0 & 0 & N_i(x) & 0 & 0 & 0 \\ 0 & 0 & 0 & N_i(x) & 0 & 0 \\ 0 & 0 & 0 & 0 & N_i(x) & 0 \\ 0 & 0 & 0 & 0 & 0 & N_i(x) \end{bmatrix} \mathbf{d}_i^e = \sum_{i=1}^{n^n} \mathbf{N}_i \mathbf{d}_i^e \tag{15}$$

where $\mathbf{d}_i^e = \left\{ u_{0i}^e \quad v_{0i}^e \quad w_{0i}^e \quad \beta_{xi}^e \quad \beta_{yi}^e \quad \beta_{zi}^e \right\}^T$ is the nodal displacement at the ith node; $N_i(\mathbf{x})$ is shape function for the ith node.

The approximation of the membrane, the bending and the shear strains of the triangular element can be written in matrix forms as follows

$$\varepsilon^e = \left[\mathbf{B}_{m1}^e \quad \mathbf{B}_{m2}^e \quad \mathbf{B}_{m3}^e \right] \mathbf{d}^e = \mathbf{B}_m^e \mathbf{d}^e \tag{16}$$

$$\kappa^e = \left[\mathbf{B}_{b1}^e \quad \mathbf{B}_{b2}^e \quad \mathbf{B}_{b3}^e \right] \mathbf{d}^e = \mathbf{B}_b^e \mathbf{d}^e \tag{17}$$

$$\gamma^e = \left[\mathbf{B}_{s1}^e \quad \mathbf{B}_{s2}^e \quad \mathbf{B}_{s3}^e \right] \mathbf{d}^e = \mathbf{B}_s^e \mathbf{d}^e \tag{18}$$

where

$$\mathbf{B}_{mi}^e = \begin{bmatrix} N_{i,x} & 0 & 0 & 0 & 0 & 0 \\ 0 & N_{i,y} & 0 & 0 & 0 & 0 \\ N_{i,y} & N_{i,x} & 0 & 0 & 0 & 0 \end{bmatrix} \tag{19}$$

$$\mathbf{B}_{bi}^e = \begin{bmatrix} 0 & 0 & 0 & N_{i,x} & 0 & 0 \\ 0 & 0 & 0 & 0 & N_{i,y} & 0 \\ 0 & 0 & 0 & N_{i,y} & N_{i,x} & 0 \end{bmatrix} \tag{20}$$

$$\mathbf{B}_{si}^e = \begin{bmatrix} 0 & 0 & N_{i,x} & N_i & 0 & 0 \\ 0 & 0 & N_{i,y} & 0 & N_i & 0 \end{bmatrix} \tag{21}$$

By substituting the discrete displacement field into Equation (9) the discretized equation for free vibration analysis can be written into matrix form such as

$$(\mathbf{K} - \omega^2 \mathbf{M})\hat{\mathbf{d}} = 0, \tag{22}$$

where ω is the natural frequency, **K** and **M** are the global stiffness and mass matrices, respectively,

$$\mathbf{K} = \sum_{e=1}^{n^e} \mathbf{T}^T \mathbf{K}^e \mathbf{T} \tag{23}$$

with

$$\mathbf{K}^e = \int_{\Omega_e} (\mathbf{B}^e)^T \hat{\mathbf{D}} \mathbf{B}^e \mathrm{d}\Omega_e \tag{24}$$

and

$$\mathbf{B}^e = \begin{bmatrix} \mathbf{B}_m^e \\ \mathbf{B}_b^e \\ \mathbf{B}_s^e \end{bmatrix}, \hat{\mathbf{D}} = \begin{bmatrix} \mathbf{D}^m & \mathbf{B} & 0 \\ \mathbf{B} & \mathbf{D}^b & 0 \\ 0 & 0 & \mathbf{C} \end{bmatrix} \tag{25}$$

$$\mathbf{M} = \sum_{e=1}^{n^e} \mathbf{T}^T \mathbf{M}^e \mathbf{T}, \tag{26}$$

$$\mathbf{M}^e = \int_{\Omega_e} \mathbf{N}^T \mathbf{m} \mathbf{N} \mathrm{d}\Omega_e \tag{27}$$

in which **T** is the transformation matrix between the local coordinate system $Oxyz$ and the global coordinate system $\hat{O}\hat{x}\hat{y}\hat{z}$ [14].

The problem of zero stiffness that appears with using the drilling degree of freedom β_z, which can cause a singularity in the global stiffness matrix when all the elements meeting at a node are coplanar. To deal with this issue, a simple modification coefficient is chosen to be 10^{-3} times the maximum diagonal value of the element stiffness matrix at the zero drilling degree of freedom to avoid the drill rotation locking [15].

3. Formulation of ES-MITC3 Finite Element Method for FGM Shells

3.1. Brief on the MITC3 Formulation

In the linear triangular MITC3, the approximated displacement field **u** is simply interpolated using the linear basic functions for membrane, deflection, and rotation without adding any new variables. Herein, the membrane and bending strains of the standard finite elements are unchanged, while the transverse shearstrains, which are modified by the mixed interpolation of tensorial components [16].

As a result, the transverse shearstrain field [8,10] is being obtained as

$$\gamma^e_{\mathrm{MITC3}} = \begin{bmatrix} \mathbf{B}^e_{s1} & \mathbf{B}^e_{s2} & \mathbf{B}^e_{s3} \end{bmatrix} \mathbf{d}^e = \mathbf{B}^e_s \mathbf{d}^e \tag{28}$$

where

$$\mathbf{B}^e_{s1} = J^{-1} \begin{bmatrix} 0 & 0 & -1 & \frac{a}{3} + \frac{d}{6} & \frac{b}{3} + \frac{c}{6} & 0 \\ 0 & 0 & -1 & \frac{d}{3} + \frac{a}{6} & \frac{c}{3} + \frac{b}{6} & 0 \end{bmatrix}, \tag{29}$$

$$\hat{\mathbf{B}}^e_{s2} = J^{-1} \begin{bmatrix} 0 & 0 & 1 & \frac{a}{2} - \frac{d}{6} & \frac{b}{2} - \frac{c}{6} & 0 \\ 0 & 0 & 0 & \frac{d}{6} & \frac{c}{6} & 0 \end{bmatrix}, \tag{30}$$

$$\hat{\mathbf{B}}^e_{s3} = J^{-1} \begin{bmatrix} 0 & 0 & 0 & \frac{a}{6} & \frac{b}{6} & 0 \\ 0 & 0 & 1 & \frac{d}{2} - \frac{a}{6} & \frac{c}{2} - \frac{b}{6} & 0 \end{bmatrix} \tag{31}$$

with

$$J^{-1} = \frac{1}{2A^e} \begin{bmatrix} c & -b \\ -d & a \end{bmatrix} \tag{32}$$

in which $a = x_2 - x_1$, $b = y_2 - y_1$, $c = y_3 - y_1$, and $d = x_3 - x_1$, as pointed out in Figure 4 and A_e is the area of the triangular element.

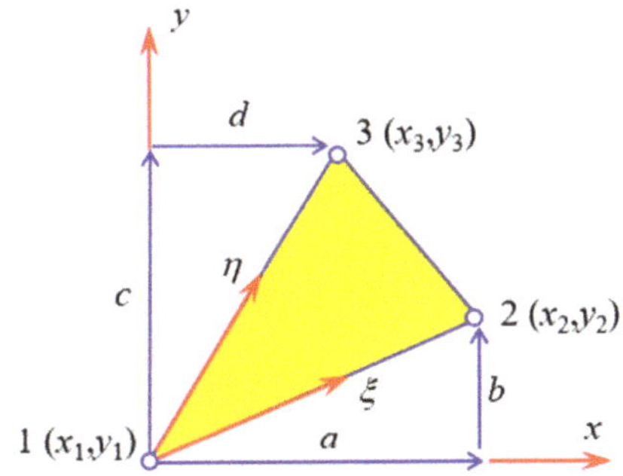

Figure 4. Three-node triangular element and local coordinates.

3.2. The ES-MITC3 Formulation

In the ES-FEM, the strains are smoothed over local smoothing domains Ω^k, the computation for stiffness matrix is no longer based on elements, but on these smoothing domains. These smoothing domains are formed based on edge of elements such as $\Omega = \cup_{k=1}^{n^k} \Omega^k$ and $\Omega^i \cap \Omega^j = \emptyset$ for $i \neq j$, in which n^k is the total number of edges of all the elements. On a curved geometry of shell models, an edge-based smoothing domain Ω^k associated with the inner edge k is created two sub-domains of two non-planar adjacent MITC3 triangular elements as shown in Figure 5. These triangular elements are defined by two local coordinate systems $O_1 x_1 y_1 z_1$ and $O_2 x_2 y_2 z_2$. In order to compute the edge-based smoothing strain $\{\}^k$ for two non-planar adjacent elements, the virtual coordinate system $\widetilde{Oxyz}$ is proposed as shown in Figure 6, whereas the $\widetilde{x}$-axis coinciding with the edge k, the $\widetilde{z}$-axis with the average direction between the $\hat{z}_1$-axis and $\hat{z}_2$-axis, and the $\widetilde{y}$-axis is given by the cross-product of the unit vectors in the $\widetilde{x}$-axis and $\widetilde{z}$-axis.

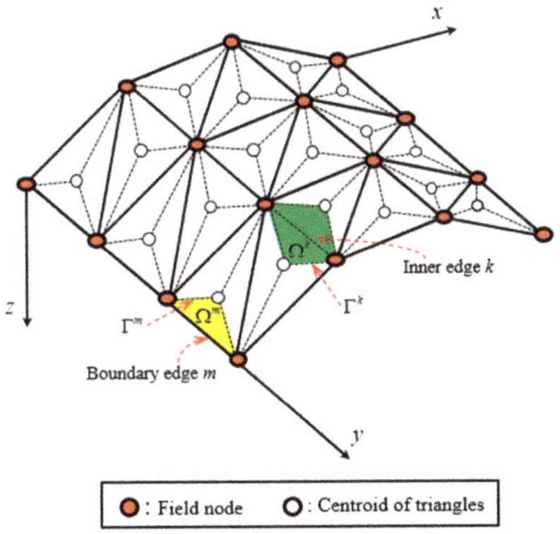

Figure 5. The smoothing domain; Ω^k is formed by triangular elements.

Hence, a smoothed membrane strain $\widetilde{\boldsymbol{\varepsilon}}^k$, a smoothed bending strain $\widetilde{\boldsymbol{\kappa}}^k$, a smoothed shearstrain $\widetilde{\boldsymbol{\gamma}}^k$ of the smoothing domain $\boldsymbol{\Omega}^k$ in the global coordinate system $\hat{O}\hat{x}\hat{y}\hat{z}$ can be derived as

$$\widetilde{\boldsymbol{\varepsilon}}^k = \sum_{j=1}^{n^{nk}} \widetilde{\mathbf{B}}_{mj}^k \mathbf{d}_j^k;\ \widetilde{\boldsymbol{\kappa}}^k = \sum_{j=1}^{n^{nk}} \widetilde{\mathbf{B}}_{bj}^k \mathbf{d}_j^k;\ \widetilde{\boldsymbol{\gamma}}^k = \sum_{j=1}^{n^{nk}} \widetilde{\mathbf{B}}_{sj}^k \mathbf{d}_j^k \tag{33}$$

where n^{nk} is the number of the neighboring nodes of edge k. $\mathbf{d}_j^k$ is the nodal degrees of freedom at the jth node of the smoothing domain Ω^k in $\hat{O}\hat{x}\hat{y}\hat{z}$. $\widetilde{\mathbf{B}}_{mj}^k$, $\widetilde{\mathbf{B}}_{bj}^k$, and $\widetilde{\mathbf{B}}_{sj}^k$ are the membrane, the bending and the MITC3 shear smoothed gradient matrices at the jth node of the smoothing domain Ω^k in the global coordinate system $\hat{O}\hat{x}\hat{y}\hat{z}$, respectively. The $\widetilde{\mathbf{B}}_{mj}^k$, $\widetilde{\mathbf{B}}_{bj}^k$ and $\widetilde{\mathbf{B}}_{sj}^k$ can be computed by

$$\widetilde{\mathbf{B}}_{mj}^k = \frac{1}{A^k}\sum_{i=1}^{n^{ek}} \frac{1}{3} A^i \boldsymbol{\Lambda}_{m1}^k \boldsymbol{\Lambda}_{m2}^i \mathbf{B}_{mj}^i \mathbf{T}_j^i \tag{34}$$

$$\widetilde{\mathbf{B}}_{bj}^k = \frac{1}{A^k}\sum_{i=1}^{n^{ek}} \frac{1}{3} A^i \boldsymbol{\Lambda}_{b1}^k \boldsymbol{\Lambda}_{b2}^i \mathbf{B}_{bj}^i \mathbf{T}_j^i \tag{35}$$

$$\widetilde{\mathbf{B}}_{sj}^k = \frac{1}{A^k}\sum_{i=1}^{n^{ek}} \frac{1}{3} A^i \boldsymbol{\Lambda}_{s1}^k \boldsymbol{\Lambda}_{s2}^i \mathbf{B}_{sj}^i \mathbf{T}_j^i \tag{36}$$

in which $\boldsymbol{\Lambda}_{m1}^k$, $\boldsymbol{\Lambda}_{b1}^k$, and $\boldsymbol{\Lambda}_{s1}^k$ are strain transformation matrices between the global coordinate system $\hat{O}\hat{x}\hat{y}\hat{z}$ and the virtual coordinate system $\widetilde{O}\widetilde{x}\widetilde{y}\widetilde{z}$, respectively; $\boldsymbol{\Lambda}_{m2}^i$, $\boldsymbol{\Lambda}_{b2}^i$ and $\boldsymbol{\Lambda}_{s2}^i$ are the strain transformation matrices between the local coordinate system $Oxyz$ of ith adjacent triangular elements and the virtual coordinate system $\widetilde{O}\widetilde{x}\widetilde{y}\widetilde{z}$, respectively; $\mathbf{T}_j^i$ is the transformation matrix between the local coordinate system $Oxyz$ at the jth node of the ith adjacent triangular element and the global coordinate system $\hat{O}\hat{x}\hat{y}\hat{z}$. More detailed information about these strain transformation matrices can be found in [14]. The area A^k of the smoothing domain Ω^k is computed by

$$A^k = \int_{\Omega^k} \mathrm{d}\Omega = \frac{1}{3}\sum_{i=1}^{n^{ek}} A^i \tag{37}$$

where n^{ek} is the number of the adjacent triangular elements in the smoothing domain Ω^k, and A^i is the area of the ith triangular element around the edge k.

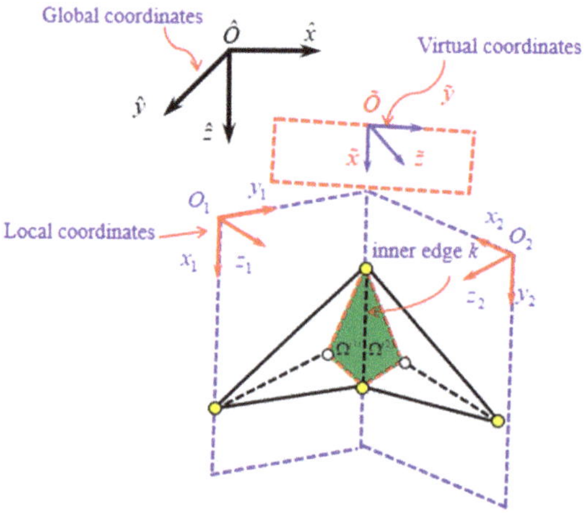

Figure 6. Global, local, and virtual coordinates.

As a result, the global stiffness matrix of the FGM shell in Equation (22) is rewritten as

$$(\widetilde{\mathbf{K}} - \omega^2\mathbf{M})\hat{\mathbf{d}} = 0, \tag{38}$$

$$\widetilde{\mathbf{K}} = \sum_{i=1}^{n^n} \widetilde{\mathbf{K}}^k \tag{39}$$

where

$$\widetilde{\mathbf{K}}^k = \int_{\Omega_k} (\widetilde{\mathbf{B}}^k)^T \hat{\mathbf{D}} \widetilde{\mathbf{B}}^k \mathrm{d}\Omega_k \tag{40}$$

with

$$\widetilde{\mathbf{B}}^k = \begin{bmatrix} \widetilde{\mathbf{B}}_m^k & \widetilde{\mathbf{B}}_b^k & \widetilde{\mathbf{B}}_s^k \end{bmatrix}^T \tag{41}$$

4. Numerical Results

In the section, several numerical examples are provided to show the performance of the ES-MITC3 element for free vibration analysis of FGM shell and results obtained are compared to those published [6, 16–19]. For convenience to the numerical comparison, the non-dimensional frequency parameters ω^* are expressed to the following equation as

$$\omega^* = \omega a^2 \sqrt{\rho_m h / D_m^*},\ D_m^* = E_m h^3 / 12(1 - v_m^3) \tag{42}$$

First, let us consider free vibration for analysis of clamped functionally graded cylindrical shell ($R_x = R,\ R_y = \infty$) with radius-to-length $R/a = 100$, $a/h = 10$. The functionally graded shell is made from Silicon nitride (Si_3N_4) and Stainless steel (SUS304), which material properties are $E_c = 322.2715$ GPa, $v_c = 0.24$, $\rho_c = 2370$ kg/m^3, $E_m = 207.7877$ GPa, $v_m = 0.31776$, and $\rho_m = 8166$ kg/m^3. The first four non dimensional frequency of the present method list in Table 1 are compared with MITC3 [16], a higher-order theory based on radial basis functions collocation including transverse normal deformation (HSDT RBFC-1) and discarding transverse normal deformation (HSDT RBFC-2) by Neves et al. [17], a higher-order theory and finite element formalation (HSDT FEM) by Pradyumna and Bandyopadhyay [6], a higher-order theory and semi-analytical method relied on Galerkin (HSDT SAG) by Yang and Shen [18], and Quasi-3D Ritz model (ED_{555}) by Fazzolari and Carrera [19]. From Table 1 we can see that this proposed method (ES-MITC3) is more accurate than other methods, such as MITC3, HSDT RBFC-1, HSDT RBFC-2, HSDT FEM and HSDT SAG. The errors are less than 3% in comparison with the exact solution ED_{555} [19]. Figure 7 shows non-dimensional frequency parameter for four first modes of clamped functionally graded cylindrical shell using various methods.

Table 1. Non-dimensional frequency parameter for clamped cylindrical functionally graded materials (FGM) shell with R/a = 100, and relative error between methods (ED_{555} [19] is fixed). Error (%) = $\frac{100\times|\text{Method}-ED_{555}[19]|}{ED_{555}[19]}$.

Mode	Method	n: 0	0.2	2	10	∞
1	ES-MITC3	75.4587	61.3587	40.9880	35.3951	33.0594
	%	0.2776	0.0298	0.3963	0.7275	0.5532
	MITC3 [16]	72.7508	59.0689	39.4771	34.0744	31.8220
	%	3.3209	3.7031	4.0679	4.4317	4.2754
	HSDT FEM [6]	72.9613	60.0269	39.1457	33.3666	32.0274
	%	3.0412	2.1413	4.8733	6.4169	3.6576
	HSDT RBFC-1 [17]	74.2634	60.0061	40.5259	35.1663	32.6108
	%	1.3108	2.1752	1.5193	1.3693	1.9026
	HSDT RBFC-2 [17]	74.5821	60.3431	40.8262	35.4229	32.8593
	%	0.8873	1.6258	0.7895	0.6496	1.1551
	HSDT SAG [18]	74.5180	57.4790	40.7500	35.8520	32.7610
	%	0.9725	6.2950	0.9747	0.5539	1.4508
	ED_{555} [19]	75.2498	61.3404	41.1511	35.6545	33.2433
2	ES-MITC3	144.4760	117.6462	78.5402	67.7320	63.3473
	%	0.6724	0.6147	0.5174	0.3138	0.4088
	MITC3 [16]	140.8063	114.5113	76.4785	65.9309	61.6559
	%	1.8847	2.0664	2.1212	2.3537	2.2722
	HSDT FEM [6]	138.5552	113.8806	74.2915	63.2869	60.5546
	%	3.4533	2.6058	4.9201	6.2695	4.0178
	HSDT RBFC-1 [17]	141.6779	114.3788	76.9725	66.6482	61.9329
	%	1.2773	2.1797	1.4889	1.2913	1.8331
	HSDT RBFC-2 [17]	142.4281	115.2134	77.6639	67.1883	62.4886
	%	0.7546	1.4660	0.6041	0.4914	0.9523
	HSDT SAG [18]	144.6630	111.7170	78.8170	69.0750	63.3140
	%	0.8027	4.4562	0.8717	2.3029	0.3560
	ED_{555} [19]	143.5110	116.9275	78.1359	67.5201	63.0894
3	ES-MITC3	145.1510	118.1985	78.9069	68.0474	63.6440
	%	1.0284	0.9602	0.8727	2.2434	0.7658
	MITC3 [16]	141.7861	115.3112	77.0122	66.3906	62.0864
	%	1.3137	1.5061	1.5494	4.6235	1.7003
	HSDT FEM [6]	138.5552	114.0266	74.3868	63.3668	60.6302
	%	3.5625	2.6033	4.9056	8.9675	4.0058
	HSDT RBFC-1 [17]	141.8485	114.5495	77.0818	66.7332	62.0082
	%	1.2702	2.1567	1.4604	4.1314	1.8241
	HSDT RBFC-2 [6]	142.6024	115.3665	77.7541	67.2689	62.5668
	%	0.7455	1.4588	0.6010	3.3618	0.9397
	HSDT SAG [18]	145.7400	112.5310	79.4070	67.5946	63.8060
	%	1.4383	3.8808	1.5121	2.8939	1.0223
	ED_{555} [19]	143.6735	117.0744	78.2242	69.6090	63.1603
4	ES-MITC3	204.0647	166.3177	111.0461	95.6539	89.5229
	%	1.1780	1.2302	1.3863	1.2447	1.2996
	MITC3 [16]	195.3261	158.8135	106.1329	91.3802	85.4901
	%	3.1547	3.3373	3.0995	3.2788	3.2637
	HSDT FEM [6]	195.5366	160.6235	104.7687	89.1970	85.1788
	%	3.0503	2.2357	4.3450	5.5896	3.6160
	HSDT RBFC-1 [17]	199.1566	160.7355	107.9484	93.3350	86.8160
	%	1.2555	2.1675	1.4419	1.2097	1.7634
	HSDT RBFC-2 [17]	200.3158	162.0337	108.9677	94.0923	87.6341
	%	0.6808	1.3773	0.5113	0.4081	0.8377
	HSDT SAG [18]	206.9920	159.8550	112.4570	98.3860	90.3700
	%	2.6294	2.7034	2.6745	4.1365	2.2581
	ED_{555} [19]	201.6888	164.2966	109.5277	94.4779	88.3744

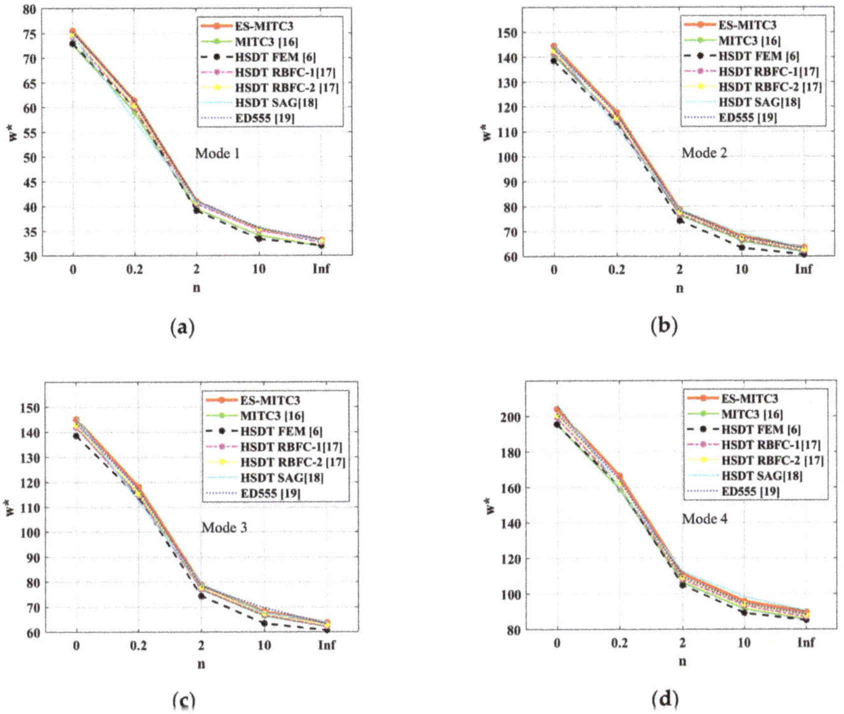

Figure 7. Non-dimensional frequency parameter for four first modes. (**a**) Mode 1; (**b**) mode 2; (**c**) mode 3; (**d**) mode 4.

Next, we investigate the first non-dimensional frequencies ω^*. of functionally graded spherical ($R_x = R_y = R$) and cylindrical shells ($R_x = R$, $R_y = \infty$) with geometric data: radius to edge R/a and a/h are varied from 5 to 50 and 10, respectively. The functionally graded shells in these studies are made from aluminum, and alumina whose material properties are $E_m = 70$, GPa, $v_m = 0.3$, $\rho_m = 2707\ \mathrm{kg/m^3}$, $E_c = 380$ GPa, $v_c = 0.3$, and $\rho_c = 3000\ \mathrm{kg/m^3}$. Again, it is seen from Tables 2–5 that the results of the present approach are very close to an HSDT RBFC-1, HSDT RBFC-2 [17], and ED_{555} [19]. Figures 8–11 show non-dimensional frequency parameter for the first mode of cylindrical FGM shell and spherical FGM shell with different n, respectively. The first six mode shapes of simply supported spherical FGM shell are illustrated in Figure 12.

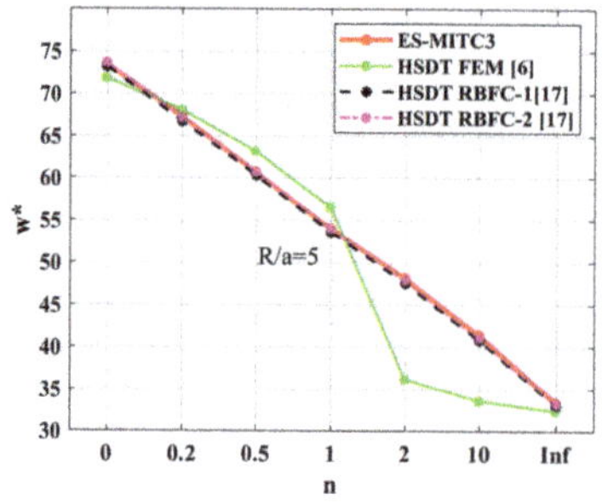

(a)

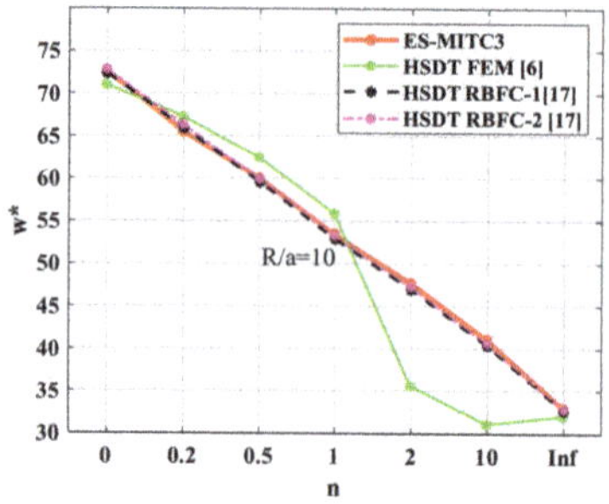

(b)

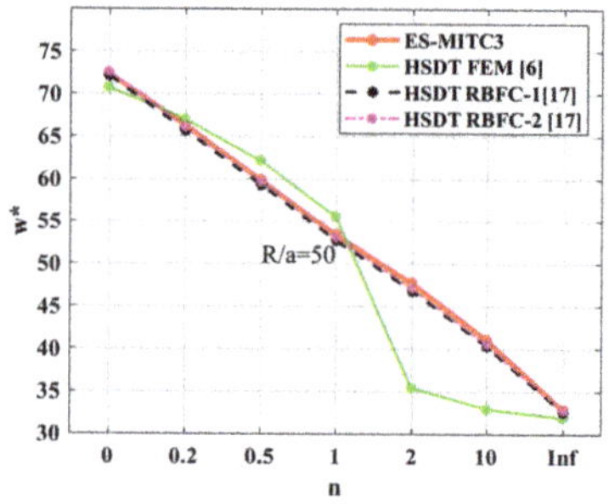

(c)

Figure 8. Non-dimensional frequency parameter for the first mode of clamped cylindrical FGM shell with different n. (**a**) R/a = 5; (**b**) R/a = 10; (**c**) R/a = 50.

Table 2. Non-dimensional frequency parameter for clamped cylindrical FGM shell with $a/h = 10$ and different R/a ratios.

$\frac{R}{a}$	Method	n						
		0	0.2	0.5	1	2	10	∞
5	ES-MITC3	73.4741	67.2928	60.6591	53.9842	48.1650	41.4718	33.4122
	HSDT FEM [6]	71.8861	68.1152	63.1896	56.5546	36.2487	33.6611	32.4802
	HSDT RBFC-1 [17]	73.1640	66.6620	60.2477	53.5430	47.5205	40.8099	33.0576
	HSDT RBFC-2 [17]	73.6436	67.1004	60.6568	53.9340	47.9060	41.0985	33.2743
10	ES-MITC3	72.6253	65.5578	60.0417	53.4874	47.7863	41.1837	33.0311
	HSDT FEM [6]	71.0394	67.3320	62.4687	55.8911	35.6633	31.1474	32.0976
	HSDT RBFC-1 [17]	72.3304	65.8808	59.5215	52.8800	46.9447	40.4145	32.6810
	HSDT RBFC-2 [17]	72.8141	66.3235	59.9353	53.2759	47.3343	40.7046	32.8995
50	ES-MITC3	72.3439	66.3519	59.9114	53.4282	47.7802	41.1529	32.9058
	HSDT FEM [6]	70.7660	67.0801	62.2380	55.6799	35.4745	32.9812	31.9741
	HSDT RBFC-1 [17]	72.0614	65.6371	59.3022	52.6864	46.7820	40.3028	32.5594
	HSDT RBFC-2 [17]	72.5465	66.0814	59.7178	53.0841	47.1726	40.5923	32.7786

Table 3. Non-dimensional frequency parameter for simply supported cylindrical FGM shell with $a/h = 10$ and different R/a ratios.

$\frac{R}{a}$	Method	n						
		0	0.2	0.5	1	2	10	∞
5	ES-MITC3	42.9913	39.3028	35.4690	31.7485	28.6106	24.7564	19.5592
	HSDT FEM [6]	42.2543	40.1621	37.2870	33.2268	27.4449	19.3892	19.0917
	HSDT RBFC-1 [17]	42.6701	38.7168	34.8768	30.9306	27.5362	24.2472	19.2796
	HSDT RBFC-2 [17]	42.7172	38.7646	34.9273	30.9865	27.5977	24.2839	19.3008
	ED_{555} [19]	42.7160	39.0642	35.0811	31.0414	27.5634	24.1245	19.3003
10	ES-MITC3	42.5231	38.9004	35.1357	31.4868	28.4168	24.6061	19.3492
	HSDT FEM [6]	41.9080	39.8472	36.9995	32.9585	27.1879	19.1562	18.9352
	HSDT RBFC-1 [17]	42.3153	38.3840	34.5672	30.6485	27.2979	24.1063	19.1193
	HSDT RBFC-2 [17]	42.3684	38.4368	34.6219	30.7077	27.3616	24.1444	19.1433
	ED_{555} [19]	42.3677	38.7377	34.7661	30.7621	27.3258	23.9848	19.1429
50	ES-MITC3	42.3669	38.7889	35.0696	31.4631	28.4233	24.5937	19.2798
	HSDT FEM [6]	41.7963	39.7465	36.9088	32.8750	27.0961	19.0809	18.8848
	HSDT RBFC-1 [17]	42.2008	38.2842	34.4809	30.5759	27.2423	24.0762	19.0675
	HSDT RBFC-2 [17]	42.2560	38.3384	34.5365	30.6355	27.3055	24.1125	19.0924
	ED_{555} [19]	42.2553	38.6391	34.6904	30.6890	27.2682	23.9515	19.0922

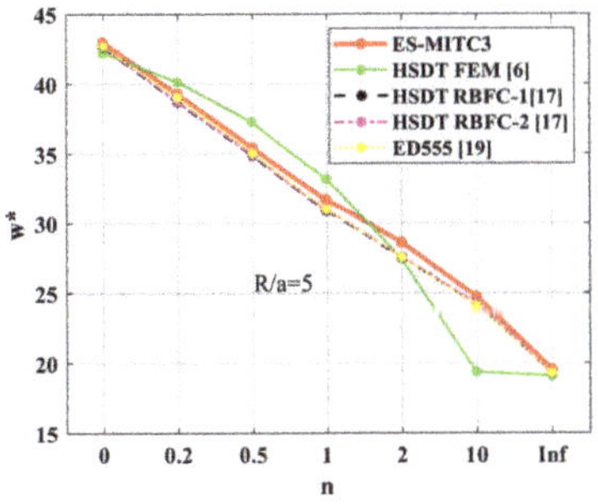

(**a**)

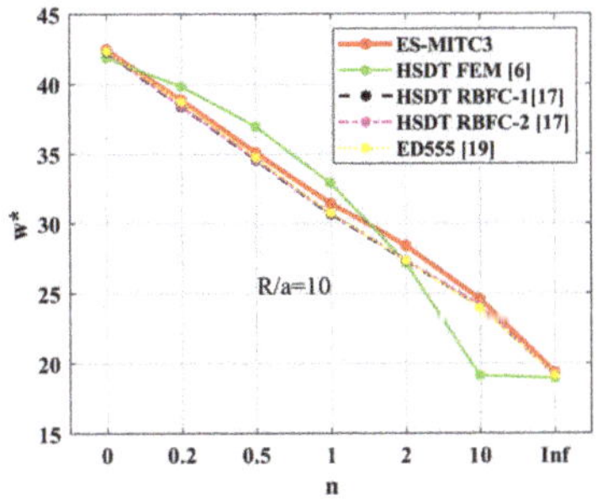

(**b**)

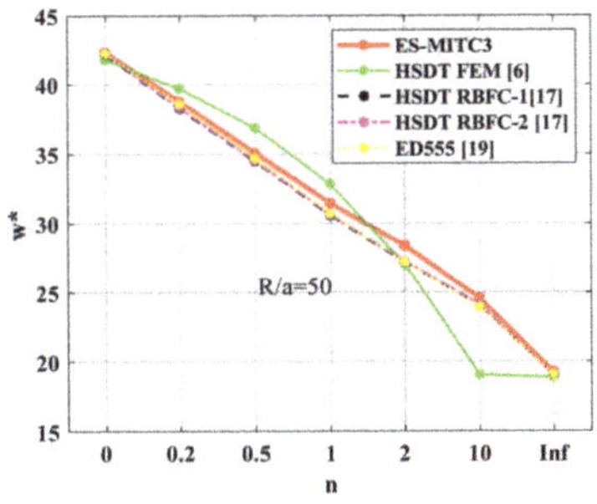

(**c**)

Figure 9. Non-dimensional frequency parameter for the first mode of simple supported cylindrical FGM shell with different n. (**a**) R/a = 5; (**b**) R/a = 10; (**c**) R/a = 50.

Table 4. Non-dimensional frequency parameter for simply supported spherical FGM shell with $a/h = 10$ and different R/a ratios.

$\frac{R}{a}$	Method	n						
		0	0.2	0.5	1	2	10	∞
5	ES-MITC3	44.4405	40.6238	36.6449	32.7529	29.4124	25.2893	20.2096
	HSDT FEM [6]	44.0073	41.7782	38.7731	34.6004	28.7459	20.4691	19.8838
	HSDT RBFC-1 [17]	44.4555	40.3936	36.4453	32.3691	28.7833	25.0772	20.0818
	HSDT RBFC-2 [17]	44.4697	40.4211	36.6004	32.4101	28.8329	25.1038	20.0927
	ED_{555} [19]	44.4671	40.7166	36.6297	32.4645	28.7996	24.9403	20.0915
10	ES-MITC3	42.9198	39.2373	35.4098	31.6957	28.5633	24.7153	19.5267
	HSDT FEM [6]	42.3579	40.2608	37.3785	33.3080	27.5110	19.4357	19.1385
	HSDT RBFC-1 [17]	42.7709	38.8074	34.9574	31.0012	27.5984	24.3034	19.3251
	HSDT RBFC-2 [17]	42.8180	38.8551	35.0080	31.0572	27.6602	24.3401	19.3464
	ED_{555} [19]	42.8169	39.1556	35.1622	31.1122	27.6258	24.1803	19.3459
50	ES-MITC3	42.4046	38.8147	35.0835	31.4662	28.4197	24.5977	19.2966
	HSDT FEM [6]	41.8145	39.7629	36.9234	32.8881	27.1085	19.0922	18.8930
	HSDT RBFC-1 [17]	42.2192	38.2988	34.4922	30.5840	27.2474	24.0791	19.0759
	HSDT RBFC-2 [17]	42.2741	38.3528	34.5478	30.6437	27.3109	24.1168	19.1006
	ED_{555} [19]	42.2735	38.6538	34.7018	30.6975	27.2741	23.9567	19.1004

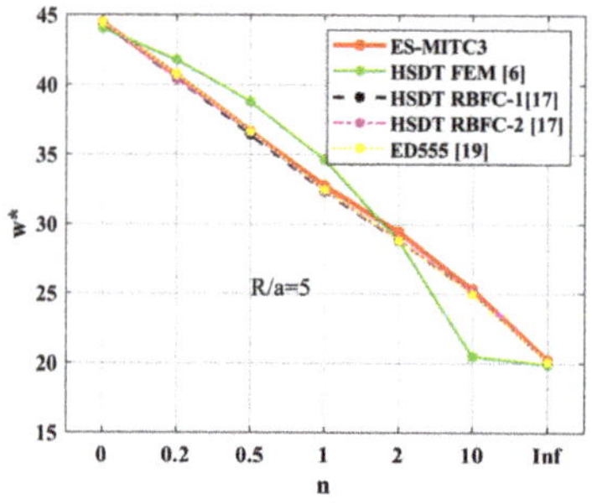

(a)

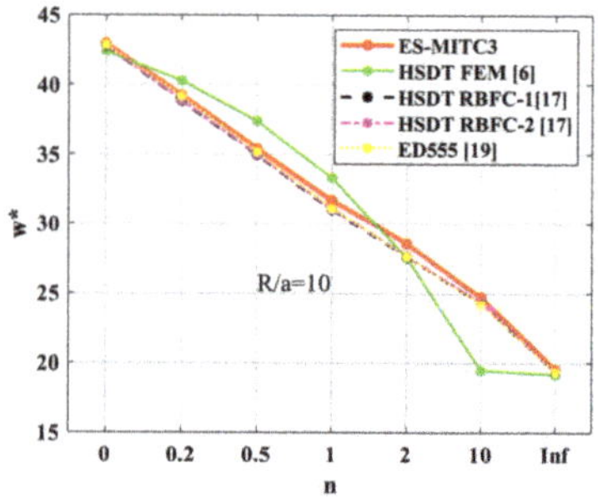

(b)

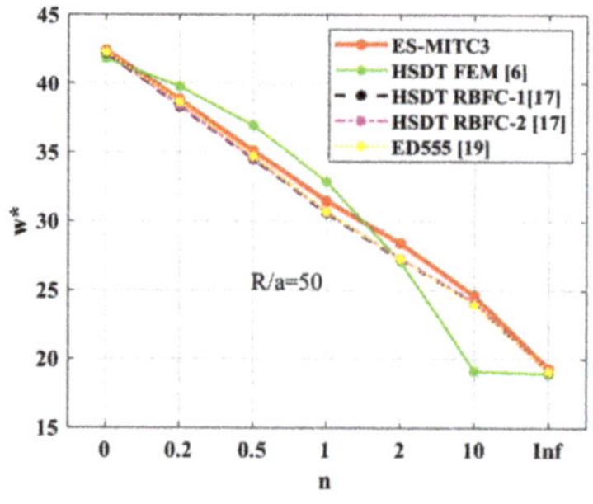

(c)

Figure 10. Non-dimensional frequency parameter for the first mode of simply supported spherical FGM shell with different n. (**a**) R/a = 5; (**b**) R/a = 10; (**c**) R/a = 50.

Table 5. Non-dimensional frequency parameter for clamped spherical FGM shell with different R/a ratios.

R/a	Method	n						
		0	0.2	0.5	1	2	10	∞
5	ES-MITC3	74.3416	68.0034	61.2122	54.3761	48.3922	41.5911	33.7972
	HSDT FEM [6]	73.5550	69.6597	64.6114	57.8619	37.3914	34.6658	33.2343
	HSDT RBFC-1 [17]	74.8207	68.2142	61.6902	54.8597	48.6656	41.6016	33.8061
	HSDT RBFC-2 [17]	75.2810	68.6329	62.0789	55.2302	49.0328	41.8796	34.0141
10	ES-MITC3	72.8831	66.7331	60.1352	53.5040	47.7428	41.1652	33.1447
	HSDT FEM [6]	71.4659	67.7257	62.8299	56.2222	35.9568	33.4057	32.2904
	HSDT RBFC-1 [17]	72.7536	66.2686	59.8745	53.1956	47.2135	40.5990	32.8722
	HSDT RBFC-2 [17]	73.2322	66.7063	60.2831	53.5864	47.5990	40.8883	33.0884
50	ES-MITC3	72.3889	66.3780	59.9190	53.4192	47.7612	41.1495	32.9258
	HSDT FEM [6]	70.7832	67.0956	62.2519	55.6923	35.4861	32.9916	31.9819
	HSDT RBFC-1 [17]	72.0784	65.6498	59.3112	52.6921	46.7849	40.3049	32.5671
	HSDT RBFC-2 [17]	72.5633	66.0938	59.7265	53.0895	47.1574	40.5946	32.7862

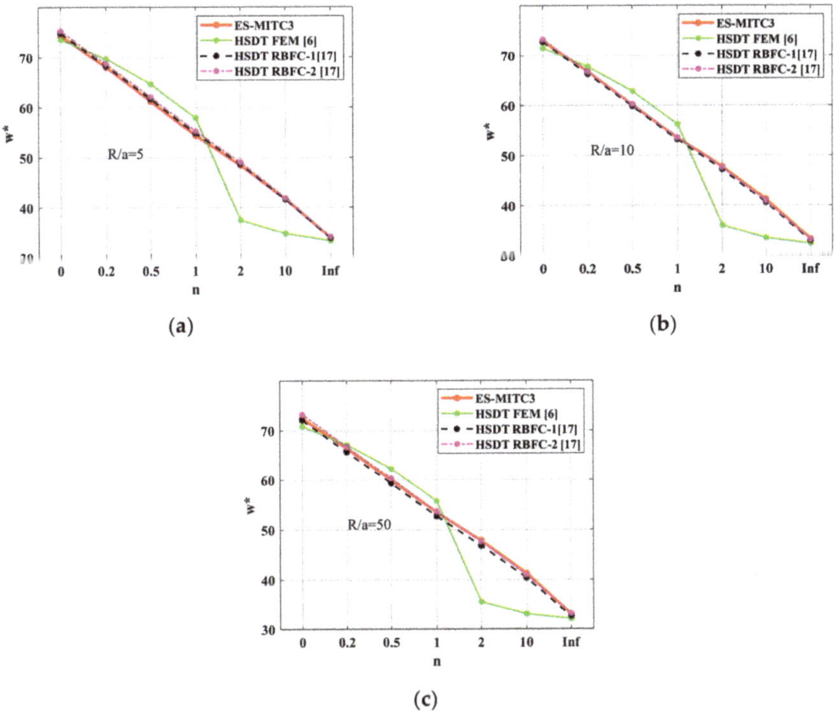

Figure 11. Non-dimensional frequency parameter for the first mode of clamped spherical FGM shell with different n. (**a**) R/a = 5; (**b**) R/a = 10; (**c**) R/a = 50.

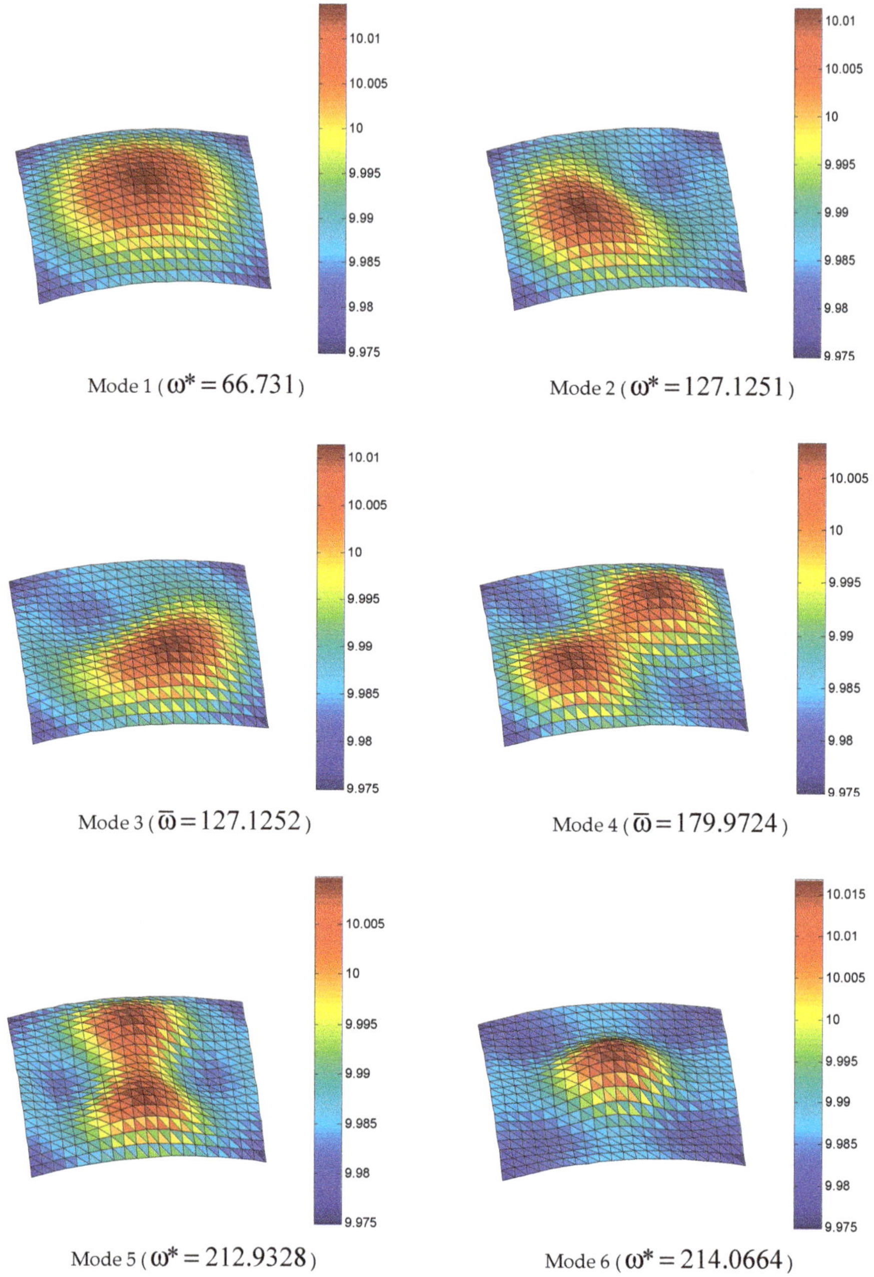

Figure 12. First six mode shapes of simply supported spherical FGM shell (R/a = 10, a/h = 10, $n = 0.2$).

For the fully clamped spherical shell, the second vibration mode shape and the third vibration mode shape are similar to each other (their natural frequencies are equal), they just have different views. Adding, the value of non-dimensional frequency of the fifth and the sixth vibration mode shape are approximatively each other. This thing is consistent with the actual symmetrical shell structures with the same boundary conditions.

5. Conclusions

In this paper, the free vibration analysis of functionally graded shells is studied by using the ES-MITC3. Herein, the stiffness matrices obtained based on the strain-smoothing technique over the smoothing domains associated with edges of MITC3 shell elements. The present approach uses a triangular element and hence much easily generated automatically, even for complicated geometries. The numerical results showed that the ES-MITC3 is a good agreement with the reference solutions, which are a requirement of high computational costs, such as ED_{555} [19] and a higher-order theory based on radial basis functions [17]. The ES-MITC3 presented herein is promising to be a simple and effective finite element method for analysis of functionally graded shells in practice.

The combination of finite element method (FEM) with an edge-based smoothed finite element method (ES-FEM) is very suitable for analyzing shell structures, especially for the complicated structures such as shell with reinforced stiffeners, reinforced nano grapheme, micro shell structures, nano shell structures, and so on. This combination allows us to calculate exactly plate and shell structures with thin thicknesses ($h = a/10^8$) due to overcoming the shear locking phenomenon.

Author Contributions: Investigation, Q.H.P.; Software, Q.H.P. and T.D.P.; Visualization, V.D.P.; Writing—original draft, V.T.D.; Writing—review & editing, H.-N.N.

Funding: This research was funded by Vietnam National Foundation for Science and Technology Development (NAFOSTED) grant number 107.02-2018.30.

Conflicts of Interest: The authors declare no conflict of interest.

References

1. Woo, J.; Meguid, S. Nonlinear analysis of functionally graded plates and shallow shells. *Int. J. Solids Struct.* **2001**, *38*, 7409–7421. [CrossRef]
2. Matsunaga, H. Free vibration and stability of functionally graded shallow shells according to a 2D higher-order deformation theory. *Compos. Struct.* **2008**, *84*, 132–146. [CrossRef]
3. Duc, N.D.; Quan, T.Q.; Luat, V.D. Nonlinear dynamic analysis and vibration of shear deformable piezoelectric FGM double curved shallow shells under damping-thermo-electro-mechanical loads. *Compos. Struct.* **2015**, *125*, 29–40. [CrossRef]
4. Bich, D.H.; Duc, N.D.; Quan, T.Q. Nonlinear vibration of imperfect eccentrically stiffened functionally graded double curved shallow shells resting on elastic foundation using the first order shear deformation theory. *Int. J. Mech. Sci.* **2014**, *80*, 16–28. [CrossRef]
5. Arciniega, R.A.; Reddy, J.N. Large deformation analysis of functionally graded shells. *Int. J. Solids Struct.* **2007**, *44*, 2036–2052. [CrossRef]
6. Pradyumna, S.; Bandyopadhyay, J.N. Free vibration analysis of functionally graded curved panels using a higher-order finite element formulation. *J. Sound Vib.* **2008**, *318*, 176–192. [CrossRef]
7. Kordkheili, S.A.H.; Naghdabadi, R. Geometrically non-linear thermoelastic analysis of functionally graded shells using finite element method. *Int. J. Numer. Methods Eng.* **2007**, *72*, 964–986. [CrossRef]
8. Chau-Dinh, T.; Nguyen-Duy, Q.; Nguyen-Xuan, H. Improvement on MITC3 plate finite element using edge-based strain smoothing enhancement for plate analysis. *Acta Mech.* **2017**, *228*, 2141–2163. [CrossRef]
9. Nguyen, T.-K.; Nguyen, V.-H.; Chau-Dinh, T.; Vo, T.P.; Nguyen-Xuan, H. Static and vibration analysis of isotropic and functionally graded sandwich plates using an edge-based MITC3 finite elements. *Compos. Part B Eng.* **2016**, *107*, 162–173. [CrossRef]
10. Pham, Q.-H.; Tran, T.-V.; Pham, T.-D.; Phan, D.-H. An Edge-Based Smoothed MITC3 (ES-MITC3) Shell Finite Element in Laminated Composite Shell Structures Analysis. *Int. J. Comput. Methods* **2017**, *15*, 1850060. [CrossRef]
11. Pham, Q.-H.; Pham, T.-D.; Trinh, Q.V.; Phan, D.-H. Geometrically nonlinear analysis of functionally graded shells using an edge-based smoothed MITC3 (ES-MITC3) finite elements. *Eng. Comput.* **2019**, 1–14. [CrossRef]
12. Pham-Tien, D.; Pham-Quoc, H.; Vu-Khac, T.; Nguyen-Van, N. Transient Analysis of Laminated Composite Shells Using an Edge-Based Smoothed Finite Element Method. In Proceedings of the International Conference on Advances in Computational Mechanics, Phu Quoc Island, Vietnam, 2–4 August 2017; pp. 1075–1094.

13. Nguyen, H.N.; Canh, T.N.; Thanh, T.T.; Ke, T.V.; Phan, V.D.; Thom, D.V. Finite Element Modelling of a Composite Shell with Shear Connectors. *Symmetry* **2019**, *11*, 527. [CrossRef]
14. Nguyen-Hoang, S.; Phung-Van, P.; Natarajan, S.; Kim, H.-G. A combined scheme of edge-based and node-based smoothed finite element methods for Reissner–Mindlin flat shells. *Eng. Comput.* **2016**, *32*, 267–284. [CrossRef]
15. Nguyen-Thoi, T.; Phung-Van, P.; Thai-Hoang, C.; Nguyen-Xuan, H. A cell-based smoothed discrete shear gap method (CS-DSG3) using triangular elements for static and free vibration analyses of shell structures. *Int. J. Mech. Sci.* **2013**, *74*, 32–45. [CrossRef]
16. Lee, P.-S.; Bathe, K.-J. Development of MITC isotropic triangular shell finite elements. *Comput. Struct.* **2004**, *82*, 945–962. [CrossRef]
17. Neves, A.; Ferreira, A.; Carrera, E.; Cinefra, M.; Roque, C.; Jorge, R.; Soares, C.; Ferreira, A.; Jorge, R.N.; Soares, C.M.M. Free vibration analysis of functionally graded shells by a higher-order shear deformation theory and radial basis functions collocation, accounting for through-the-thickness deformations. *Eur. J. Mech. A/Solids* **2013**, *37*, 24–34. [CrossRef]
18. Yang, J.; Shen, H.-S. Free vibration and parametric resonance of shear deformable functionally graded cylindrical panels. *J. Sound Vib.* **2003**, *261*, 871–893. [CrossRef]
19. Fazzolari, F.A.; Carrera, E. Refined hierarchical kinematics quasi-3D Ritz models for free vibration analysis of doubly curved FGM shells and sandwich shells with FGM core. *J. Sound Vib.* **2014**, *333*, 1485–1508. [CrossRef]

symmetry

MDPI

Article

Finite Element Modelling of a Composite Shell with Shear Connectors

Hoang-Nam Nguyen [1], Tran Ngoc Canh [2], Tran Trung Thanh [3], Tran Van Ke [3], Van-Duc Phan [4,*] and Do Van Thom [3,*]

1 Modeling Evolutionary Algorithms Simulation and Artificial Intelligence, Faculty of Electrical & Electronics Engineering, Ton Duc Thang University, Ho Chi Minh City 700000, Vietnam; nguyenhoangnam@tdtu.edu.vn
2 Department of Mechanics, Tran Dai Nghia University, Ho Chi Minh City 700000, Vietnam; canhvhp@gmail.com
3 Faculty of Mechanical Engineering, Le Quy Don Technical University, Hanoi City 100000, Vietnam; tranthanh0212@gmail.com (T.T.T.); tranke92@gmail.com (T.V.K.)
4 Center of Excellence for Automation and Precision Mechanical Engineering, Nguyen Tat Thanh University, Ho Chi Minh City 700000, Vietnam
* Correspondence: pvduc@ntt.edu.vn (V.-D.P.); thom.dovan.mta@gmail.com (D.V.T.)

Received: 1 March 2019; Accepted: 29 March 2019; Published: 11 April 2019

Abstract: A three-layer composite shell with shear connectors is made of three shell layers with one another connected by stubs at the contact surfaces. These layers can have similar or different geometrical and physical properties with the assumption that they always contact and have relative movement in the working process. Due to these characteristics, they are used widely in many engineering applications, such as ship manufacturing and production, aerospace technologies, transportation, and so on. However, there are not many studies on these types of structures. This paper is based on the first-order shear deformation Mindlin plate theory and finite element method (FEM) to establish the oscillator equations of the shell structure under dynamic load. The authors construct the calculation program in the MATLAB environment and verify the accuracy of the established program. Based on this approach, we study the effects of some of the geometrical and physical parameters on the dynamic responses of the shell.

Keywords: three-layer composite shell; Mindlin plate theory; finite element method; force vibration

1. Introduction

Nowadays, along with a strong development of science and technology, there are many new advanced materials appeared, for instance, composite materials, functionally graded materials (FGM), piezoelectric materials, and so on. The studying on dynamic responses of these new materials has been reached great achievements and attracted numerous scientists all over the world. Moreover, the idea of merging these different materials is considered by engineers to make new structures in order to have specific purposes. For example, the combining of a concrete structure and a steel structure has a lighter weight than a normal concrete structure. Hence, these new types of structures are applied extensively in civil techniques, aerospace, and army vehicles. In this structure, the connecting stub is attached to contact different layers in order to create the compatibility of the horizontal displacement among layers, and it plays an important role in working process of the structure.

For multilayered beams, recently, the Newark's model [1] is considered by many experts such as He et al. [2], Xu and Wang [3,4]. They took into account the shear strain when calculating by using Timoshenko beam theory. Nguyen [5] studied the linear dynamic problems. Silva et al. [6], Schnabl et al. [7] and Nguyen and co-workers [8,9] employed the finite element method (FEM) and analytical method in order to examine linear static analysis of multilayered beams. Huang [10], and Shen [11] studied the linear dynamic response, too. For the nonlinear free vibration can be seen in [12] of Arvin and Bakhtiari-Nejad.

In addition to the Timoshenko beam theory (TBT), the higher-order beam theory (HBT) is also considered, in which The dynamic problem is carried out by Chakrabarti in [13] with FEM. Chakrabarti and colleagues [14,15] analyzed a static problem for two-layer composite beams. The higher-order beam theory (HBT) overcomes a part of the effect due to the shear locking coefficient caused. Otherwise, Subramanian [16] constructed an element based on a displacement field to study the free vibration of the multilayered beam. Li et al. [17] conducted a free vibration analysis by employing the hyperbolic shear deformation theory. Vo and Thai [18] studied static multilayered beams with the improved higher-order beam theory of Shimpi.

In general, most higher-order beam theories (HBT), including higher-order beam theory of Reddy tend to neglect the horizontal deformation of multilayered beams. According to the Kant's opinion, the horizontal stress of sub-layer is caused by the pressure can reduce the dimension from multilayered beam model to the plane stress model. To obtain this thing, Kant [19,20] employed both the higher-order beam theory (HBT) and the horizontal displacement theory by considering approximate displacements in two ways. Thus, he established the mixed two-layer beam with sub-layers, which abides by the higher-order beam theory of Kant proposed by the weak form for the buckling analysis.

A three-dimensional fracture plasticity based on finite element model (FEM) are developed by Yan and coworkers [21] to carry out the ultimate strength respones of SCS sandwich structure under concentrated loads. The static behaviors of beams with different types of cross-section, such as square, C-shaped, and bridge-like sections, were investigated in Carrena's study [22] by assuming that the displacement field is expanded in terms of generic functions, which is the Unified Formulation by Carrera (CUF) [23]. Similarly to mentioned methods, Cinefra et al. [24] used MITC9 shell elements to explore the mechanical behavior of laminated composite plates and shells. Muresan and coworkers [25] examined the study on the stability of thin walled prismatic bars based on the Generalized Beam Theory (GBT), which is an efficient approach developed by Schardt [26]. Yu et al. [27] employed the Variational Asymptotic Beam Section Analysis (VABS) for mechanical behavior of various cross-sections such as elliptic and triangular sections. In [28], we used first-order shear deformation theory to analysis of triple-layer composite plates with layers connected by shear connectors subjected to moving load. Ansari Sadrabadi et al. [29] used analytical methods to investigate a thick-walled cylindrical tube made of a functionally graded material (FGM) and undergoing thermomechanical loads.

For multilayered plate and shell composite structures, there have been many published papers, including static problems, dynamic problems, linear, and nonlinear problems, and so on. However, for the multilayered structure with shear connectors, there are not many papers yet. Based on above mentioned papers, the authors are about to construct the relations of mechanical behavior and the oscillator equation of the multilayered shell. We also study several geometrical and physical parameters, the loading, etc., which effect on the vibration of the shell.

The body of this paper is divided into five main sections. Section 1 is the general introduction. We present finite formulations of free vibration and forced vibration analysis of three-layer composite shell with shear connectors in Section 2. The numerical results of vibration and forced vibrations are discussed in Sections 3 and 4. Section 5 gives some major conclusions.

2. Finite Element Formulations

2.1. Equation of Motion of the Shell Element

Consider a three-layer composite shell with shear connectors as shown in Figure 1.

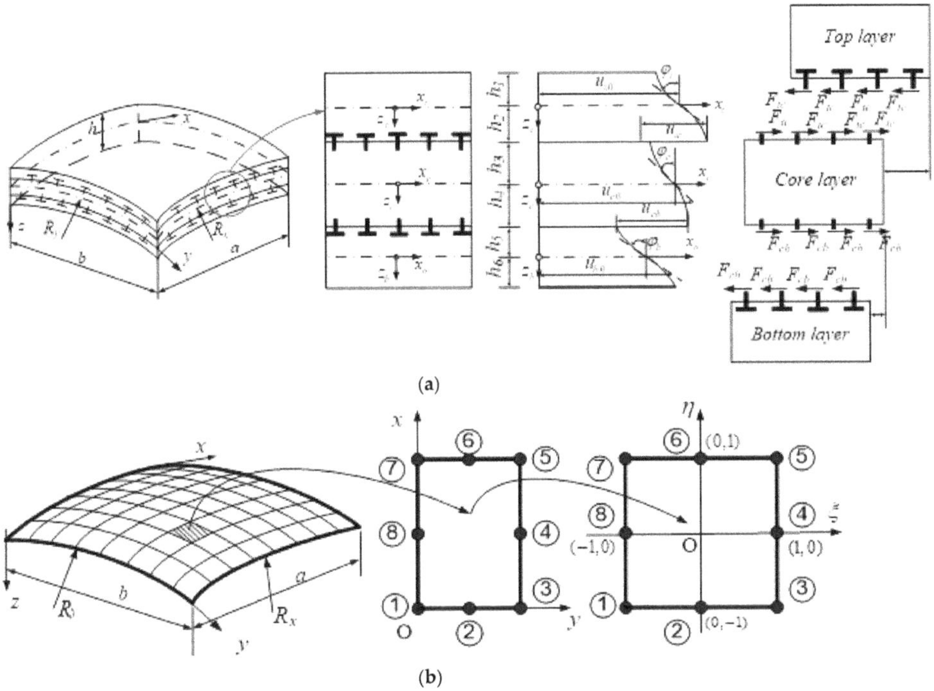

Figure 1. The model of three-layer composite shell with shear connectors, (**a**) shell model with shear connectors, and (**b**) finite element model.

The composite shell consists of three layers, including the top layer (t), the bottom layer (b) and the middle layer (c); these layers are connected with one another by shear connectors, and they can be made of the same materials or different materials. These three layers can slide relatively with one another at the contact surfaces, and there is no delamination phenomenon at all. All three layers of the shell are set in the local coordinates $Oxyt$, $Oxyc$, and $Oxyb$, respectively. The total thickness of the shell is divided into six small part $h_1, h_2, h_3, h_4, h_5, h_6$ as shown in Figure 1; u_{t0}, u_{c0}, and u_{b0} represent displacements in x direction; v_{b0} represents the displacement in y direction at the neutral surface of each layer.

According to Mindlin plate theory, displacements u, v, w at a point (x_k, y_k, z_k) of layer k are expressed as follows:

$$\begin{cases} u_k = u_{k0}(x_k, y_k) + z_k\varphi_k(x_k, y_k) \\ v_k = v_{k0}(x_k, y_k) + z_k\psi_k(x_k, y_k) \quad (k = t, c, b) \\ w_k = w(x_k, y_k) \end{cases} \tag{1}$$

where φ_k and ψ_k are the transverse normal rotations of the x_k and y_k directions.

The relative movements among the contact surfaces are defined by the following equations

For the layer t and layer c we have

$$\begin{cases} u_{tc} = u_t(x_t, y_t, h_2) - u_c(x_c, y_c, -h_3) \\ v_{tc} = v_t(x_t, y_t, h_2) - v_c(x_c, y_c, -h_3) \end{cases} \tag{2}$$

And for layer c and layer b we have:

$$\begin{cases} u_{cb} = u_c(x_c, y_c, h_4) - u_b(x_b, y_b, -h_5) \\ v_{cb} = v_c(x_c, y_c, h_4) - v_b(x_b, y_b, -h_5) \end{cases} \tag{3}$$

Note that at the contact surfaces, we have:

$$\begin{cases} z_t = h_2; z_c = -h_3 \\ z_c = h_4; z_b = -h_5 \end{cases} \tag{4}$$

with $h_4 = h_3 = \frac{h_c}{2}$.

From Equations (1)–(4), we get:

$$\begin{cases} u_{tc} = u_{t0} - u_{c0} + h_2\varphi_t + h_3\varphi_c \\ v_{tc} = v_{t0} - v_{c0} + h_2\psi_t + h_3\psi_c \end{cases} \tag{5}$$

$$\begin{cases} u_{cb} = u_{c0} - u_{b0} + h_4\varphi_c + h_5\varphi_b \\ v_{cb} = v_{c0} - v_{b0} + h_4\psi_c + h_5\psi_b \end{cases} \tag{6}$$

The relation between strain and displacement of each layer is expressed as follows

For the layer k, we have:

$$\begin{aligned}
\varepsilon_{kx} &= \frac{\partial u_k}{\partial x} = \frac{\partial u_{k0}}{\partial x} + \frac{w_0}{R_x} + z_k\frac{\partial \varphi_k}{\partial x}; \\
\varepsilon_{ky} &= \frac{\partial v_k}{\partial y} = \frac{\partial v_{k0}}{\partial y} + \frac{w_0}{R_y} + z_k\frac{\partial \psi_k}{\partial y}; \\
\gamma_{kxy} &= \frac{\partial v_k}{\partial x} + \frac{\partial u_k}{\partial y} = \frac{\partial u_{k0}}{\partial y} + \frac{\partial v_{k0}}{\partial x} + \frac{2w_0}{R_{xy}} + z_k\left(\frac{\partial \varphi_k}{\partial y} + \frac{\partial \psi_k}{\partial x} + \frac{1}{2}\left(\frac{1}{R_y} - \frac{1}{R_x}\right)\left(\frac{\partial v_{k0}}{\partial x} - \frac{\partial u_{k0}}{\partial y}\right)\right); \\
\gamma_{kxz} &= \frac{\partial w_0}{\partial x} + \frac{\partial u_k}{\partial z_k} = \frac{\partial w_0}{\partial x} + \varphi_k - \frac{u_{k0}}{R_x} - \frac{v_{k0}}{R_{xy}}; \\
\gamma_{kyz} &= \frac{\partial w_0}{\partial y} + \frac{\partial v_k}{\partial z_k} = \frac{\partial w_0}{\partial y} + \psi_k - \frac{u_{k0}}{R_{xy}} - \frac{v_{k0}}{R_y};
\end{aligned} \tag{7}$$

We can rewrite in a matrix form as follow

$$\boldsymbol{\varepsilon}_k = \begin{Bmatrix} \varepsilon_{kx} \\ \varepsilon_{ky} \\ \gamma_{kxy} \end{Bmatrix} = \boldsymbol{\varepsilon}_k^0 + z_k\boldsymbol{\kappa}_k;\ \boldsymbol{\gamma}_k = \begin{Bmatrix} \gamma_{kyz} \\ \gamma_{kzx} \end{Bmatrix} \tag{8}$$

in which

$$\begin{aligned}
\boldsymbol{\varepsilon}_k^0 &= \begin{Bmatrix} \varepsilon_{kx}^0 \\ \varepsilon_{ky}^0 \\ \gamma_{kxy}^0 \end{Bmatrix} = \begin{Bmatrix} \frac{\partial u_{k0}}{\partial x} + \frac{w_0}{R_x} \\ \frac{\partial v_{k0}}{\partial y} + \frac{w_0}{R_y} \\ \left(\frac{\partial u_{k0}}{\partial y} + \frac{\partial v_{k0}}{\partial x}\right) + \frac{2w_0}{R_{xy}} \end{Bmatrix};\ \boldsymbol{\kappa}_k = \begin{Bmatrix} \kappa_{kx} \\ \kappa_{ky} \\ \kappa_{kxy} \end{Bmatrix} = \begin{Bmatrix} \frac{\partial \varphi_k}{\partial x} \\ \frac{\partial \psi_k}{\partial y} \\ \frac{\partial \varphi_k}{\partial y} + \frac{\partial \psi_k}{\partial x} + \frac{1}{2}\left(\frac{1}{R_y} - \frac{1}{R_x}\right)\left(\frac{\partial v_{k0}}{\partial x} - \frac{\partial u_{k0}}{\partial y}\right) \end{Bmatrix} \\
\boldsymbol{\gamma}_k &= \begin{Bmatrix} \gamma_{kxz} \\ \gamma_{kyz} \end{Bmatrix} = \begin{Bmatrix} -\frac{u_{k0}}{R_x} - \frac{v_{k0}}{R_{xy}} + \frac{\partial w_0}{\partial x} + \varphi_k \\ -\frac{u_{k0}}{R_{xy}} - \frac{v_{k0}}{R_y} + \frac{\partial w_0}{\partial y} + \psi_k \end{Bmatrix}
\end{aligned} \tag{9}$$

The relation between stress and strain of layer k is expressed as followIs necessary bild?

$$\boldsymbol{\sigma}_k = \mathbf{D}_k\boldsymbol{\varepsilon}_k;\ \boldsymbol{\tau}_k = \frac{5}{6}\mathbf{G}_k\boldsymbol{\gamma}_k \tag{10}$$

in which $\mathbf{D}_k$, $\mathbf{G}_k$ are the bending rigidity and shear rigidity of layer k, respectively, and ν_k is the Poisson ratio of layer k.

$$\mathbf{D}_k = \frac{E_k}{1-v^2}\begin{bmatrix} 1 & \nu_k & 0 \\ \nu_k & 1 & 0 \\ 0 & 0 & (1-\nu_k)/2 \end{bmatrix};\ \mathbf{G}_k = \frac{E_k}{2(1+\nu_k)}\begin{bmatrix} 1 & 0 \\ 0 & 1 \end{bmatrix} \tag{11}$$

In this work, the thickness of the shell is thin or medium ($h = \frac{a}{100} \div \frac{a}{10}$, a is short edge), we employ the 8-node isoparametric element, each node has 13 degrees of freedom, three layers have the same displacement in the z-direction (Figure 2), the degree of freedom of node i is $\mathbf{q}_e^i$ and the total degree of freedom of the shell element $\mathbf{q}_e$ is defined as follow.

$$\mathbf{q}_e^i = \{\ u_{t0i}\quad v_{t0i}\quad \varphi_{ti}\quad \psi_{ti}\quad u_{c0i}\quad v_{c0i}\quad \varphi_{ci}\quad \psi_{ci}\quad u_{b0i}\quad v_{b0i}\quad \varphi_{bi}\quad \psi_{bi}\quad w\}^{\mathrm{T}};\ i = 1 \div 8. \tag{12}$$

$$\mathbf{q}_e = \{\ \mathbf{q}_e^1\quad \mathbf{q}_e^2\quad \mathbf{q}_e^3\quad \mathbf{q}_e^4\quad \mathbf{q}_e^5\quad \mathbf{q}_e^6\quad \mathbf{q}_e^7\quad \mathbf{q}_e^8\ \}^{\mathrm{T}} \tag{13}$$

$$\begin{array}{c} u_{k0} = \sum\limits_{i=1}^{8} N_i(\xi,\eta)u_{k0i};\ v_{k0} = \sum\limits_{i=1}^{8} N_i(\xi,\eta)v_{k0i} \\ \varphi_k = \sum\limits_{i=1}^{8} N_i(\xi,\eta)\varphi_{ki};\ \psi_k = \sum\limits_{i=1}^{8} N_i(\xi,\eta)\psi_{ki};\ w = \sum\limits_{i=1}^{8} N_i(\zeta,\eta)w_i \end{array} \quad (k = t, c, b) \tag{14}$$

in which N_i ($i = 1 \div 8$) can be defined as in [28].

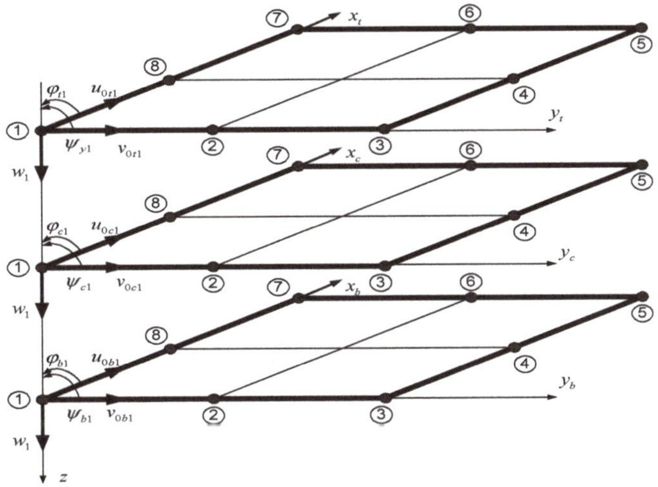

Figure 2. Degrees of freedom of the node in the eight-node shell element.

By substituting in the expression for verifying displacement of element we have:

$$\left\{ \begin{array}{c} \varepsilon_k = \left(\mathbf{B}_k^0 + z_k\mathbf{B}_k^1\right)\mathbf{q}_e \\ \gamma_k = \mathbf{S}_k\mathbf{q}_e \end{array} \right. \quad (k = t, c, b) \tag{15}$$

in which $\mathbf{B}_k^0$; $\mathbf{B}_k^1$; $\mathbf{S}_k$ are defined as follows

$$\begin{array}{l} \mathbf{B}_k^0 = \left[\ \mathbf{B}_{k1}^0\quad \mathbf{B}_{k2}^0\quad \mathbf{B}_{k3}^0\quad \mathbf{B}_{k4}^0\quad \mathbf{B}_{k5}^0\quad \mathbf{B}_{k6}^0\quad \mathbf{B}_{k7}^0\quad \mathbf{B}_{k8}^0\ \right]; \\ \mathbf{B}_k^1 = \left[\ \mathbf{B}_{k1}^1\quad \mathbf{B}_{k2}^1\quad \mathbf{B}_{k3}^1\quad \mathbf{B}_{k4}^1\quad \mathbf{B}_{k5}^1\quad \mathbf{B}_{k6}^1\quad \mathbf{B}_{k7}^1\quad \mathbf{B}_{k8}^1\ \right]; \\ \mathbf{S}_k = \left[\ \mathbf{S}_{k1}\quad \mathbf{S}_{k2}\quad \mathbf{S}_{k3}\quad \mathbf{S}_{k4}\quad \mathbf{S}_{k5}\quad \mathbf{S}_{k6}\quad \mathbf{S}_{k7}\quad \mathbf{S}_{k8}\ \right]; \end{array} \tag{16}$$

where $\mathbf{B}_{ki}^0$, $\mathbf{B}_{ki}^1$ and $\mathbf{S}_{ki}$ can be found in Appendix A

The elastic force of connector stub per unit length is defined by the following equations.

For layer t and c we have:

$$\mathbf{F}_e^{tc} = \left\{ \begin{array}{c} F_{eu} \\ F_{ev} \end{array} \right\}_{ct} = k_{tc}\left[\begin{array}{cc} 1 & 0 \\ 0 & 1 \end{array} \right]\left\{ \begin{array}{c} u_{tc} \\ v_{tc} \end{array} \right\} = \mathbf{K}_e^{tc}\mathbf{q}_e^{tc} \tag{17}$$

With

$$\mathbf{q}_e^{tc} = \left\{ \begin{array}{c} u_{tc} \\ v_{tc} \end{array} \right\} = \left[\begin{array}{c} u_{t0} + h_2\varphi_t - u_{c0} + h_3\varphi_c \\ v_{t0} + h_2\psi_t - v_{c0} + h_3\psi_c \end{array} \right] = \mathbf{N}_{tc}\mathbf{q}_e = \sum_{i=1}^{8} (N_{tc})_i q_e^i \tag{18}$$

in which

$$(\mathbf{N}_{tc})_i = \begin{bmatrix} N_i & 0 & h_2N_i & 0 & -N_i & 0 & h_3N_i & 0 & 0\,0\,0\,0\,0 \\ 0 & N_i & 0 & h_2N_i & 0 & -N_i & 0 & h_3N_i & 0\,0\,0\,0\,0 \end{bmatrix} \tag{19}$$

For layer c and b we have

$$\mathbf{F}_e^{cb} = \begin{Bmatrix} F_{eu} \\ F_{ev} \end{Bmatrix}_{cb} = k_{cb}\begin{bmatrix} 1 & 0 \\ 0 & 1 \end{bmatrix}\begin{Bmatrix} u_{cb} \\ v_{cb} \end{Bmatrix} = \mathbf{K}_e^{cb}\mathbf{q}_e^{cb} \tag{20}$$

with

$$\mathbf{q}_e^{cb} = \begin{Bmatrix} u_{cb} \\ v_{cb} \end{Bmatrix} = \begin{bmatrix} u_{c0} + h_4\varphi_c - u_{b0} + h_5\varphi_b \\ v_{c0} + h_4\psi_c - v_{b0} + h_5\psi_b \end{bmatrix} = \mathbf{N}_{cb}\mathbf{q}_e = \sum_{i=1}^{8} (N_{cb})_i q_e^i \tag{21}$$

in which

$$(\mathbf{N}_{cb})_i = \begin{bmatrix} 0\,0\,0\,0\,N_i & 0 & h_4N_i & 0 & -N_i & 0 & h_5N_i & 0 & 0 \\ 0\,0\,0\,0\,0 & N_i & 0 & h_4N_i & 0 & -N_i & 0 & h_5N_i & 0 \end{bmatrix} \tag{22}$$

Here, k_{tc} and k_{cb} are the shear resistance coefficients of the connector stub per unit length. To obtain the dynamic equation we employ the weak form for each element, we get:

$$\begin{aligned} &\sum_{k=t,c,b}\int_{V_k} \delta\dot{\mathbf{q}}_k^{\mathrm{T}}\rho_k\dot{\mathbf{q}}_k dV_k + \sum_{k=t,c,b}\int_{V_k} \delta\varepsilon_k^{\mathrm{T}}\boldsymbol{\sigma}_k dV_k + \tfrac{5}{6}\sum_{k=t,c,b}\int_{V_k} \delta\gamma_k^{\mathrm{T}}\boldsymbol{\tau}_k dV_k + \sum_{k=tc,cb}\int_{A_k} \delta\left(\mathbf{q}_e^k\right)^{\mathrm{T}}\mathbf{F}_e^k dA_k \\ &-\delta\mathbf{q}_e^{\mathrm{T}}\int_{A_t} \mathbf{N}_w p(t) dA_t = 0 \end{aligned} \tag{23}$$

By substituting Equations (1), (15), (17), and (20) into Equation (23), we obtain the dynamic equation of the shell element as follows:

$$\mathbf{M}_e\ddot{\mathbf{q}}_e + \mathbf{K}_e\mathbf{q}_e = \mathbf{F}_e(t) \tag{24}$$

with

$$\begin{aligned} \mathbf{K}_{e(104\mathrm{x}104)} =& \sum_{k=c,s,a}\int_{A_k}\left(\mathbf{B}_k^0\right)^{\mathrm{T}}\mathbf{D}_{k0}\mathbf{B}_k^0 dA_k + \sum_{k=c,s,a}\int_{A_k}\left(\mathbf{B}_k^0\right)^{\mathrm{T}}\mathbf{D}_{k1}\mathbf{B}_k^1 dA_k + \\ &+ \sum_{k=t,c,b}\int_{A_k}\left(\mathbf{B}_k^1\right)^{\mathrm{T}}\mathbf{D}_{k1}\mathbf{B}_k^0 dA_k + \sum_{k=t,c,b}\int_{A_k}\left(\mathbf{B}_k^1\right)^{\mathrm{T}}\mathbf{D}_{k2}\mathbf{B}_k^1 dA_k + \tfrac{5}{6}\sum_{k=t,c,b}\int_{A_k}\mathbf{S}_k^{T}\mathbf{G}_k\mathbf{S}_k dA_k \\ &+ \int_{A_{tc}}\mathbf{N}_{tc}^{\mathrm{T}}\mathbf{K}_{tc}^e\mathbf{N}_{tc} dA_{tc} + \int_{A_{cb}}\mathbf{N}_{cb}^{\mathrm{T}}\mathbf{K}_{cb}^e\mathbf{N}_{cb} dA_{cb} \end{aligned} \tag{25}$$

in which

$$(\mathbf{D}_{k0};\ \mathbf{D}_{k1};\ \mathbf{D}_{k2}) = \int_{-h_k/2}^{h_k/2}\left(1;\ z_k;\ z_k^2\right)\mathbf{D}_k\, dz_k;\ \mathbf{H}_k = \int_{-h_k/2}^{h_k/2}\mathbf{G}_k\, dz_k\ (k = t, c, b) \tag{26}$$

$$\mathbf{M}_{e(104\mathrm{x}104)} = \sum_{k=t,c,b}\int_{A_k}\int_{-h_k/2}^{h_k/2}\mathbf{L}_k^{\mathrm{T}}\rho_k\mathbf{L}_k dz_k dA_k \tag{27}$$

where $\mathbf{L}_k$ can be seen in Appendix B

$$\mathbf{F}_e(t)_{(104\mathrm{x}1)} = \int_{A_t} p(t)\mathbf{N}_{\mathrm{w}}^{\mathrm{T}} dA_t \tag{28}$$

in which

$$\mathbf{N}_w = \begin{bmatrix} \mathbf{N}_{w1} & \mathbf{N}_{w2} & \mathbf{N}_{w3} & \mathbf{N}_{w4} & \mathbf{N}_{w5} & \mathbf{N}_{w6} & \mathbf{N}_{w7} & \mathbf{N}_{w8} \end{bmatrix} \tag{29}$$

with

$$\mathbf{N}_{wj} = \begin{bmatrix} 0 & 0 & 0 & 0 & 0 & 0 & 0 & 00000 & N_j \end{bmatrix} \tag{30}$$

In the case of taking into account the structural damping, we have the force vibration equation of the shell element as follows:

$$\mathbf{M}_e\ddot{\mathbf{q}}_e + \mathbf{C}_e\dot{\mathbf{q}}_e + \mathbf{K}_e\mathbf{q}_e = \mathbf{F}_e(t) \tag{31}$$

in which $\mathbf{C}_e = \alpha\mathbf{M}_e + \beta\mathbf{K}_e$ and $\alpha,\ \beta$ are Rayleigh drag coefficients defined in [30,31].

2.2. The Differential Equation of Vibration

From the differential equation of vibration of the shell element (Equation (31)), we obtain the differential equation of forced vibration of three-layer composite shell as follows:

$$\mathbf{M}\ddot{\mathbf{q}} + \mathbf{C}\mathbf{q} + \mathbf{K}\mathbf{q} = \mathbf{F}(t) \tag{32}$$

in which **M**, **C**, **K**, $\mathbf{F}(t)$ are the global mass matrix, the global structural damping matrix, the global stiffness matrix and the global load matrix, respectively. These matrices and vectors are assembled from the element matrices and vectors, correspondingly. They are linear differential equations, which have the right-hand side depending on time. In order to solve these equations, we use the Newmark-beta method [31]. The program is coded in the MATLAB (MathWorks, Natick, MA, USA) environment with the following algorithm flowchart of Newmark as shown Figure 3.

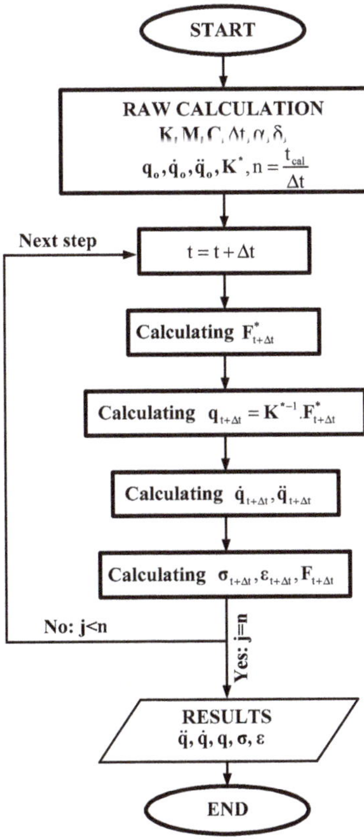

Figure 3. Algorithm flowchart of Newmark solving the dynamic response problem of the shell.

For the free vibration analysis, the natural frequencies can be obtained by solving the equation:

$$\mathbf{M}\ddot{\mathbf{q}} + \mathbf{K}\mathbf{q} = 0 \tag{33}$$

or in another form:

$$\left(\mathbf{K} - \omega^2\mathbf{M}\right)\mathbf{q} = 0 \tag{34}$$

where ω is the natural frequency.

Flowchart of Newmark-beta method [31]

Step 1: Determine the first conditions:

$$\mathbf{q}(0) = \mathbf{q}_0; \dot{\mathbf{q}}(0) = \dot{\mathbf{q}}_0 \tag{35}$$

From the first conditions, we obtain:

$$\ddot{\mathbf{q}}_0 = \mathbf{M}_0^{-1}\left(\mathbf{F}_0 - \mathbf{K}_0\mathbf{q}_0 - \mathbf{C}_0\dot{\mathbf{q}}_0\right) \tag{36}$$

Step 2: By approximating $\ddot{\mathbf{q}}_{t+\Delta t}, \dot{\mathbf{q}}_{t+\Delta t}$ by $\mathbf{q}_{t+\Delta t}$, we have

$$\begin{aligned} \ddot{\mathbf{q}}_{t+\Delta t} &= a_0\left(\mathbf{q}_{t+\Delta t} - \mathbf{q}_t\right) - a_2\dot{\mathbf{q}}_t - a_3\ddot{\mathbf{q}}_t \\ \dot{\mathbf{q}}_{t+\Delta t} &= \dot{\mathbf{q}}_t + a_6\ddot{\mathbf{q}}_t + a_7\ddot{\mathbf{q}}_{t+\Delta t} \end{aligned} \tag{37}$$

where:

$$\begin{aligned} a_0 = \frac{2}{\gamma\Delta t^2};\ a_1 = \frac{2\alpha}{\gamma\Delta t};\ a_2 = \frac{2}{\gamma\Delta t};\ a_3 = \frac{1}{\gamma} - 1;\ a_4 = \frac{2\alpha}{\gamma} - 1; \\ a_5 = \left(\frac{\alpha}{\gamma} - 1\right)\Delta t;\ a_6 = (1-\alpha)\Delta t;\ a_7 = \alpha\Delta t. \end{aligned} \tag{38}$$

in which α, γ are defined by the assumption that the acceleration varies in each calculating step, the author selects the linear law for the varying of acceleration:

$$\ddot{\mathbf{q}}(\tau) = \ddot{\mathbf{q}}_t + \frac{\tau}{\Delta t}\left(\ddot{\mathbf{q}}_{t+\Delta t} - \ddot{\mathbf{q}}_t\right) \text{ with } t \le \tau \le t + \Delta t \text{ then } \alpha = \frac{1}{2}; \gamma = \frac{1}{3}. \tag{39}$$

The condition to stabilize the roots:

$$\Delta t \le \frac{1}{\sqrt{2}\omega_{\max}}\frac{1}{\sqrt{\alpha - \gamma}} \text{ or } \frac{\Delta t}{T_{\min}} \le \frac{1}{2\pi\sqrt{2}}\frac{1}{\sqrt{\alpha - \gamma}} \tag{40}$$

Step 3: Calculating the stiffness matrix and the nodal force vector:

$$\mathbf{K}^* = \overline{\mathbf{K}} + a_0\overline{\mathbf{M}} + a_1\overline{\mathbf{C}} \tag{41}$$

$$\mathbf{F}^* = \overline{\mathbf{F}}_{t+\Delta t} + \overline{\mathbf{M}}\left(a_0\mathbf{q}_t + a_2\dot{\mathbf{q}}_t + a_3\ddot{\mathbf{q}}_t\right) + \overline{\mathbf{C}}\left(a_1\mathbf{q}_t + a_4\dot{\mathbf{q}}_t + a_5\ddot{\mathbf{q}}_t\right) \tag{42}$$

Step 4: Determining nodal displacement vector $\mathbf{q}_{t+\Delta t}$:

$$\mathbf{K}^*{}_{t+\Delta t}\mathbf{q}_{t+\Delta t} = \mathbf{F}^*{}_{t+\Delta t} \tag{43}$$

$$\Rightarrow \mathbf{q}_{t+\Delta t} = \left(\mathbf{K}^*{}_{t+\Delta t}\right)^{-1}\mathbf{F}^*{}_{t+\Delta t} \tag{44}$$

repeating the loop until the time runs out.

3. Numerical Results of Free Vibration Analysis of Three-Layer Composite Shells with Shear Connectors

3.1. Accuracy Studies

Consider a double-curved composite shell ($0^0/90^0/0^0$) with geometrical parameters $a = b$, radii $R_x = R_y = R$, thickness h; physical parameters $E_1 = 25E_2$, $G_{23} = 0.2E_2$, $G_{13} = G_{12} = 0.5E_2$, the Poisson's ratio $\nu_{12} = 0.25$, and the specific weight ρ. In this case, the shear coefficient of the stub has a very large value, and this time the three-layer composite shell becomes a normal composite shell without any relative movements. We examine the convergence of the algorithm with different meshes and the comparative results of the first non-dimensional free vibration $\overline{\omega} = \omega_1 \frac{a^2}{h}\sqrt{\frac{\rho}{E_2}}$ with Reddy [32] are shown in Table 1.

Table 1. The first non-dimensional fundamental frequencies with different meshes.

a/h = 100		This Work					Reddy [32]
Meshes		4 × 4	6 × 6	8 × 8	10 × 10	12 × 12	
	1	126.430	126.135	126.145	126.145	126.145	125.99
	2	68.489	68.095	68.065	68.065	68.065	68.075
	3	47.432	47.316	47.369	47.369	47.369	47.265
R/a	4	36.989	36.975	37.083	37.083	37.083	36.971
	5	31.188	30.908	31.030	31.030	31.030	30.993
	10	20.313	20.350	20.332	20.332	20.332	20.347
	10^{30}	15.174	5.151	15.1457	15.1457	15.1457	15.183
a/h = 10		This Work					Reddy [32]
Meshes		4 × 4	6 × 6	8 × 8	10 × 10	12 × 12	
	1	16.3576	16.3272	16.3226	16.3226	16.3226	16.115
	2	12.9939	12.9811	12.978	12.978	12.978	13.382
	3	12.1582	12.1500	12.1488	12.1488	12.1488	12.731
R/a	4	11.8418	11.8354	11.8343	11.8343	11.8343	12.487
	5	11.6905	11.6851	11.6843	11.6843	11.6843	12.372
	10	11.4843	11.4799	11.4791	11.4791	11.4791	12.215
	10^{30}	11.4141	11.4102	11.4095	11.4095	11.4095	12.165

From Table 1 we can see clearly that, in comparison between this work and the analytical method [32], we have good agreement, demonstrating that our proposed theory and program are verified for the free vibration problem and convergence is guaranteed with 8 × 8 meshes.

3.2. Effects of Some Parameters on Free Vibration of the Shell

We now consider a three-layer composite shell with geometrical parameters: length a is constant, width b, radii $R_x = R_y = R$, the total thickness h, the thickness of the middle layer h_c, the thicknesses of the other layers $h_t = h_b$ ($h_1 = h_2 = h_t/2$, $h_3 = h_4 = h_c/2$, $h_5 = h_6 = h_b/2$); physical parameters: the elastic modulus $E_c = 70$ GPa, $E_t = E_b = 200$ GPa, the Poisson's ratio $\nu_t = \nu_c = \nu_b = 0.3$, the specific weight $\rho_c = 2300$ kg/m^3,$\rho_t = \rho_b = 7800$ kg/m^3, the shear coefficient of the shear connector $k_{tc} = k_c = k_s$, and the shell structure is fully supported. We conduct an investigation into the first non-dimensional free vibration of the shell with non-dimensional frequencies as defined by:

$$\overline{\omega} = \omega_1 \frac{a^2}{h_0}\sqrt{\frac{\rho_t}{E_t}} \text{with } h_0 = \frac{a}{50} \quad (45)$$

3.2.1. Effect of Thickness h

Firstly, to examine the effect of length-to-high ratio a/h, a is fixed, we consider three cases with a/h = 75, 60, 50, 25, 10 (respectively). In each case, the radius-to-length ratio R/a changes from 1 to 10 as we can see in Table 2, $b = a$, $h_c/h_t = 2$, and the shear coefficient of stub $k_s = 50$ MPa. The results are presented in Table 2.

Table 2. Effect of thickness h on non-dimensional fundamental frequencies.

R/a	$a/h = 75$	$a/h = 60$	$a/h = 50$	$a/h = 25$	$a/h = 10$
1	48.9232	48.9329	48.9447	49.0591	49.8350
2	25.2821	25.3022	25.3267	25.5643	27.1440
3	16.9857	17.0161	17.0531	17.4099	19.6929
4	12.7982	12.8388	12.8881	13.3594	16.2390
5	10.2817	10.3324	10.3938	10.9744	14.3496
6	8.6065	8.6671	8.7403	9.4243	13.2070
7	7.4135	7.4838	7.5686	8.3498	12.4663
8	6.5225	6.6024	6.6983	7.5705	11.9605
9	5.8331	5.9222	6.0291	6.9856	11.6008
10	5.2847	5.3830	5.5004	6.5351	11.3363

Table 2 demonstrates that when the length-to-high ratio a/h decreases, that means the stiffness of the structure is enhanced, correspondingly with each case of the radius-to-length ratios R/a, the non-dimensional fundamental frequency increases.

3.2.2. Effect of the h_c/h_t Ratio ($h_t = h_b$)

Next, in order to study the effect of the h_c/h_t ratio, we consider five cases with h_c/h_t, respectively given values from 2, 4, 8, 20, 30, $b = a$ (a is fixed), the total thickness $h = a/50$, and the shear coefficient of the stub $k_s = 50$ MPa. The numerical results are shown in Table 3.

Table 3. Effect of h_c/h_t ratio on non-dimensional fundamental frequencies.

R/a	$h_c/h_t = 2$	$h_c/h_t = 4$	$h_c/h_t = 8$	$h_c/h_t = 20$	$h_c/h_t = 30$
1	48.9447	49.6002	50.3956	51.3623	51.6856
2	25.3267	25.7143	26.2246	26.8744	27.0959
3	17.0531	17.3679	17.8202	18.4208	18.6285
4	12.8881	13.1816	13.6348	14.2529	14.4682
5	10.3938	10.6864	11.1625	11.8208	12.0502
6	8.7403	9.0417	9.5499	10.2558	10.5011
7	7.5686	7.8839	8.4277	9.1822	9.4432
8	6.6983	7.0301	7.6105	8.4117	8.6871
9	2.6219	6.3787	6.9948	7.8393	8.1278
10	2.5746	5.8683	6.5186	7.4026	7.7027

Table 3 gives us a discussion that when increasing h_c/h_t ratio, and for h is constant, that means the thickness of the middle layer increases, correspondingly each case of R/a ratios, the non-dimensional fundamental frequency increases. This shows that when the thickness of the shell is constant, h_c/h_t increases, thus, the non-dimensional fundamental frequency increases.

3.2.3. Effect of the Length-to-Width Ratio a/b

In this small section, we continually evaluate the effect of the length-to-width ratio a/b (a is fixed), and we meditate three cases by letting $a/b = 0.5, 0.75, 1, 1.75$, and 2, respectively. The total thickness of the shell $h = a/50$, $h_c/h_t = 2$, the radius-to-length R/a also varies from 1 to 10, as we can see in Table 4, and the shear coefficient of stub $k_s = 50$ MPa. The numerical results are tabulated in Table 4.

In Table 4 we can see obviously that, with each value of radius-to-length R/a, if the length-to-width a/b increases, the non-dimensional fundamental frequency also increases, correspondingly. This interesting point demonstrates that the stiffness of the structure is enhanced.

Table 4. Effect of length-to-width ratio *a/b* on non-dimensional fundamental frequencies.

R/a	*a/b* = 0.5	*a/b* = 0.75	*a/b* = 1	*a/b* = 1.5	*a/b* = 2
1	47.6987	48.3535	48.9447	49.8156	50.4172
2	25.1170	25.2201	25.3267	25.5626	25.9090
3	16.9310	16.9833	17.0531	17.2818	17.7252
4	12.7690	12.8155	12.8881	13.1627	13.7259
5	10.2602	10.3104	10.3938	10.7236	11.4034
6	8.5872	8.6438	8.7403	9.1263	9.9146
7	7.3943	7.4584	7.5686	8.0095	8.8965
8	6.5025	6.5743	6.6983	7.1919	8.1678
9	5.8117	5.8914	6.0291	6.5726	7.6279
10	5.2618	5.3494	5.5004	6.0909	7.2169

3.2.4. Effect of the Shear Coefficient of Stub k_s

Finally, in this section, to examine how the shear coefficient of the stub affects the non-dimensional fundamental frequencies of this structure, we consider three cases of shear coefficient as in Table 5, and $a = b$, $h = a/50$, $h_c/h_t = 2$, $E_c = 70$ GPa is fixed. The numerical results are shown in this table.

Table 5. Effect of shear coefficient of the stub k_s on non-dimensional fundamental frequencies.

R/a	$\frac{k_s}{E_c}$=1.45×10^{-5}	$\frac{k_s}{E_c}$=1.45×10^{-2}	$\frac{k_s}{E_c}$=1.45×10^{0}	$\frac{k_s}{E_c}$=1.45×10^{2}	$\frac{k_s}{E_c}$=1.45×10^{5}
1	48.9437	48.9457	49.0792	49.3528	49.3628
2	25.3247	25.3288	25.6047	26.1739	26.1918
3	17.0500	17.0562	17.4689	18.3090	18.3350
4	12.8840	12.8922	13.4361	14.5187	14.5518
5	10.3887	10.3989	11.0674	12.3634	12.4024
6	8.7342	8.7463	9.5324	11.0130	11.0568
7	7.5615	7.5756	8.4716	10.1104	10.1582
8	6.6904	6.7062	7.7046	9.4781	9.5291
9	6.0202	6.0379	7.1307	9.0186	9.0722
10	5.4907	5.5100	6.6899	8.6749	8.7306

In Table 5 we can see clearly that, with one value of radius-to-length *R/a*, when the shear coefficient of stub increases, the non-dimensional fundamental frequency of the structure get larger. This explains that the increasing of the shear coefficient removes the slip among layers, leading to an increase of the total stiffness of the shell structure.

4. Numerical Results of Forced Vibration Analysis of Three-Layer Composite Shells with Shear Connectors

4.1. Accuracy Studies

Considerign that a fully-clamped square plate with parameters can be found in [33], $a = b = 1m$, $h/a = 10$. Material properties are the elastic modulus $E = 30$ GPa, the Poisson's ratio $\nu = 0.3$, $\rho = 2800$ kg/m^3. The structure is subjected to distribution sudden load $p_0 = 10^4$ Pa. The non-dimensional displacement is calculated by the formula $w^* = \frac{100Eh^3}{12p_0a^4(1-\nu^2)}w_0$. By taking the shear coefficient and radii of the shell as very large, the comparative deflection of the centroid point of the plate between our work and [33] is shown in Figure 4, where the integral time is 5 ms, and the acting time of load is 2 ms.

We can see from Figure 4 that the deflection of the centroid point of the plate is compared to [33] is similar both shape and value. This proves that our program is verified.

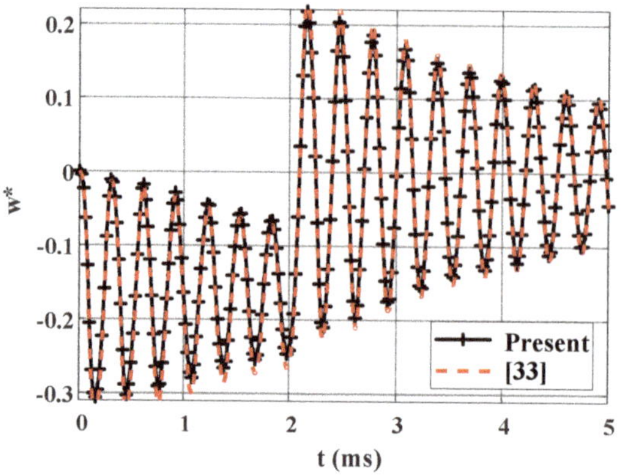

Figure 4. The deflection of the centroid point of the plate overtime.

4.2. Effect of Some Parameters on the Forced Vibration of the Shell

Now, to study effects of some parameters on forced vibration of shell, we consider a three-layer composite shell with geometrical parameters: length a =1 m, width b, thickness h, radii of the shell $R_x = R_y = R$, the thickness of middle layer h_c, the thickness of other layers $h_t = h_b$. Material properties are the elastic modulus E_c = 8 GPa, $E_t = E_b$ = 12 GPa, the Poisson's ratio $\nu_t = \nu_c = \nu_b = 0.2$, the specific weight $\rho_c = 700$ kg/m^3, $\rho_t = \rho_b = 2300$ kg/m^3, and the shear coefficient of stub $k_{tc} = k_{cb} = k_s$. The shell is fully clamped with the uniform load $p(t)$ varying overtime acting perpendicularly on the shell surface.

$$p(t) = \Delta P_\Phi.F(t); F(t) = \begin{cases} 1 - \frac{t}{\tau_{hd}}(0 \le t \le \tau_{hd}) \\ 0 \text{ otherwise} \end{cases} \text{with} \begin{cases} \Delta P_\Phi = 0.20679.10^6 \text{ N/m}^2 \\ \tau_{hd} = 0.028\, s \end{cases} \tag{46}$$

The non-dimensional deflection and velocity of the centroid point over time are given as follows:

$$\begin{aligned} w^* &= \frac{100{h_0}^3 E_t}{\Delta P_\Phi a^4} w\left(\frac{a}{2}, \frac{b}{2}\right);\ v^* = \frac{T{h_0}^3 E_t}{\Delta P_\Phi a^4} v\left(\frac{a}{2}, \frac{b}{2}\right) \\ u_c^* &= \frac{10{h_0}^3 E_c}{Mga^2(1-\nu_c^2)} u_c\left(\frac{a}{2}, \frac{b}{2}, -\frac{h_c}{2}\right); v_c^* = \frac{10{h_0}^3 E_c}{Mga^2(1-\nu_c^2)} v_c\left(\frac{a}{2}, \frac{b}{2}, -\frac{h_c}{2}\right) \text{with } h_0 = \frac{a}{50};\ T = 0.15(s) \end{aligned} \tag{47}$$

where $w\left(\frac{a}{2}, \frac{b}{2}\right)$ and $v\left(\frac{a}{2}, \frac{b}{2}\right)$ are the deflection and velocity of the centroid point of the shell.

4.2.1. Effect of the Length-to-High Ratio a/h

In this first small section, we study the effect of the length-to-high ratio a/h. We consider a shell with geometrical parameters $a = b$ (a is fixed), $h_c/h_t = 2$, $R/a = 6$, and a/h gets value 75, 60, 50, 40 and 25, respectively, the shear coefficient of stub k_s = 50 MPa. The non-dimensional deflection and velocity of the centroid point of the shell are presented in Figure 5 and the maximum value is shown in Table 6.

From Figure 4 and Table 6 we can see that when reducing the value of a/h ratio, this means the thickness of the shell gets thicker, the non-dimensional deflection and velocity of the centroid point overtime decrease. This is a good agreement, the reason is when the thickness of the shell increases, the stiffness of the shell obviously becomes higher.

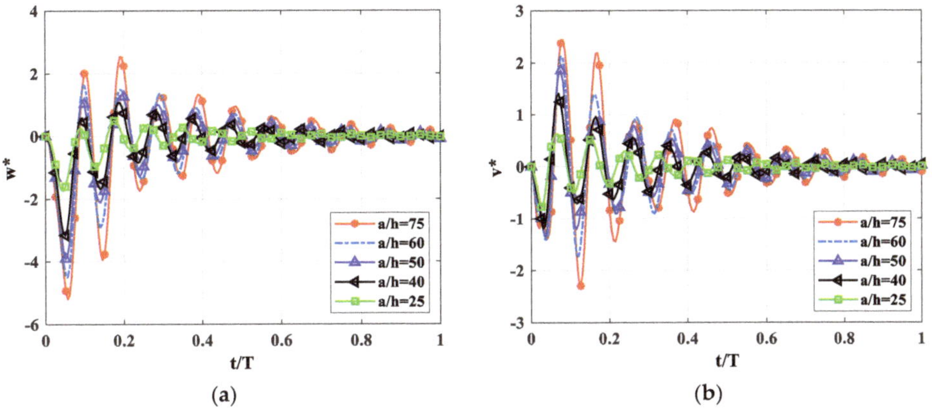

Figure 5. Effect of length-to-high ratio *a/h* on the non-dimensional deflection and velocity of the centroid point. (**a**) Nondimensional deflection w* versus time; and (**b**) nondimensional velocity v* versus time.

Table 6. Effect of length-to-high ratio a/h the non-dimensional deflection and velocity of the centroid point.

Maximum Values	$a/h = 75$	$a/h = 60$	$a/h = 50$	$a/h = 40$	$a/h = 25$
w^*_{max}	5.1866	4.5001	3.9020	3.1534	1.7250
v^*_{max}	2.2997	1.7296	1.3653	1.1890	0.7686

4.2.2. Effect of the h_c/h_t Ratio ($h_t = h_b$)

Next, to investigate the effect of h_c/h_t ratio, we dissect the shell with geometrical parameters $a = b$ (a is fixed); $h = a/50$, the value of h_c/h_t ratio is given as 2, 6, 8, 10, 20, and 30, $R/a = 6$, and the shear coefficient of stub $k_s = 50$ (MPa). The non-dimensional deflection and velocity of the centroid point of the shell are shown in Figure 6, the maximum value is listed in Table 7.

We can see in Figure 6 and Table 7 that when the h_c/h_t ratio increases (h is constant), the thickness of the middle layer increases in comparison to the other layers, and the non-dimensional deflection and velocity of the centroid point overtime decreases quickly in a range from 2–20. In a range from 20–30 the non-dimensional deflection and velocity of the centroid point overtime are almost not changed. The reason is explained that when the value of the h_c/h_t ratio increases, the structure can reduce the ability to oscillate, and the middle layer becomes "softer", so that it imbues the vibration better than a homogeneous shell with same geometrical and physical parameters. For this particular problem, we should select the value of h_c/h_t ratio in a range from 20–30.

Table 7. Effect of h_c/h_t ratio on the non-dimensional deflection and velocity of the centroid point.

Maximum Values	$h_c/h_t = 2$	$h_c/h_t = 6$	$h_c/h_t = 8$	$h_c/h_t = 10$	$h_c/h_t = 20$	$h_c/h_t = 30$
w^*_{max}	2.4284	2.3267	2.2499	2.1578	2.0313	1.9787
$u_c^{*max} \times 10^{-5}$	1.1921	1.8455	1.9861	1.990	2.1182	2.1605
$v_c^{*max} \times 10^{-6}$	5.1472	4.8345	5.0843	5.4432	6.9771	8.6697
v^*_{max}	1.1221	1.4232	1.4658	1.4525	1.4479	1.4879

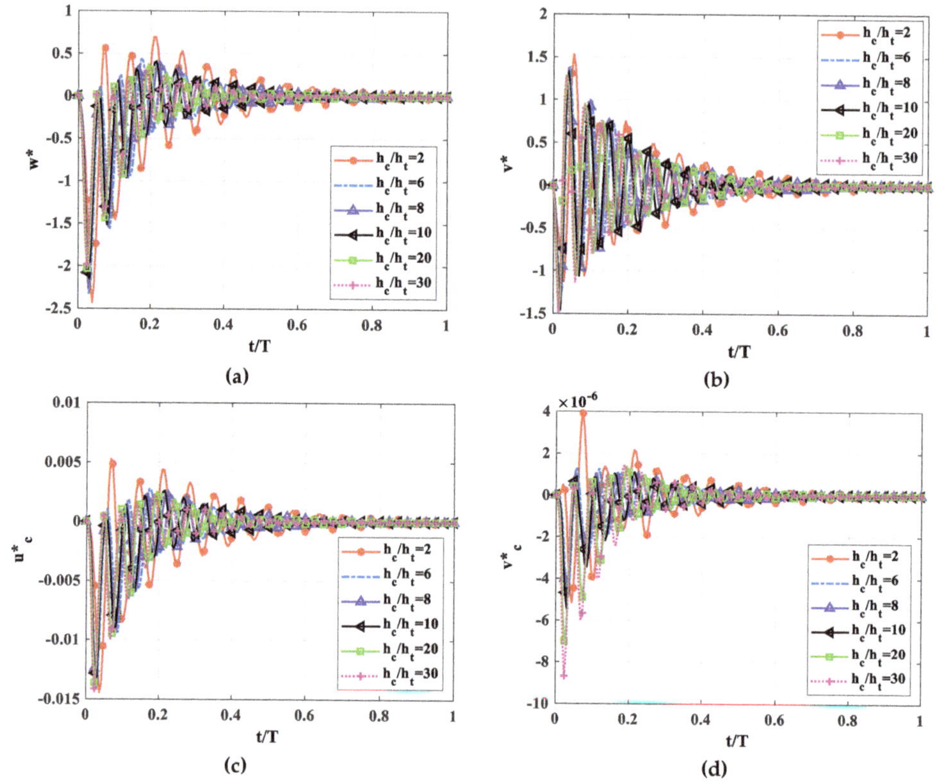

Figure 6. Effect of h_c/h_t ratio on the non-dimensional deflection and velocity of the centroid point. (**a**) Nondimensional velocity w* versus time; (**b**) Nondimensional velocity v* versus time; (**c**) Nondimensional deflection u_c^* versus time; (**d**) Nondimensional deflection v_c^* versus time.

4.2.3. Effect of the Length-to-Width Ratio *a/b*

We examine the effect of length-to-width ratio a/b on the non-dimensional deflection and velocity of the centroid point of the shell with a is fixed, a/b gets from 0.5, 1, 1.5, 2. The geometrical parameters are $h = a/50$, $h_c/h_t = 2$, $R/a = 6$, and the shear coefficient of stub $k_s = 50$ MPa. The numerical results of non-dimensional deflection and velocity of the centroid point of the shell are shown in Figure 7, and the maximum value is listed in the following Table 8.

Table 8. Effect of the length-to-width ratio *a/b* on the non-dimensional deflection and velocity of the centroid point.

Maximum Values	$a/b = 0.5$	$a/b = 0.75$	$a/b = 1$	$a/b = 1.5$	$a/b = 2$
w^*_{max}	3.1051	3.5580	3.9020	3.6051	3.0260
v^*_{max}	1.0248	1.2536	1.3653	1.3871	1.1701

Now we can see in Figure 7 and Table 8, when increasing the a/b ratio, the non-dimensional deflection and velocity of the centroid point overtime decrease. This demonstrates that the stiffness of the shell gets larger, especially when the a/b ratio equals 2. This can be understood obviously that as the shape of structure gets smaller, with the same boundary condition and other parameters, the structure will become stronger.

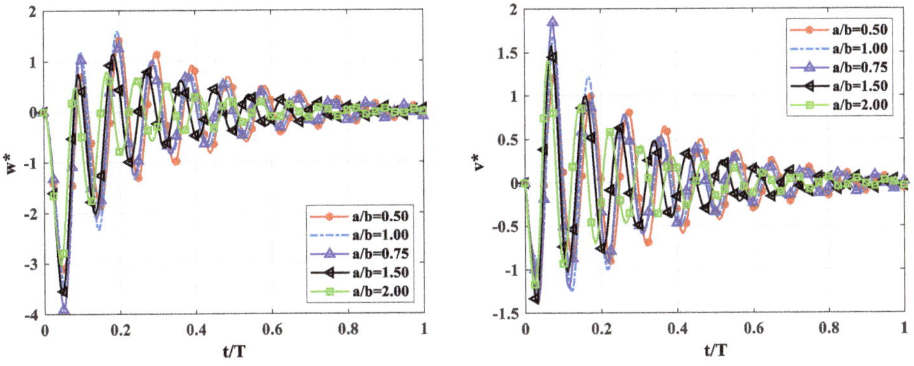

(a) (b)

Figure 7. Effect of length-to-width ratio *a/b* on the non-dimensional deflection and velocity of the centroid point. (**a**) Nondimensional deflection w* versus time; and (**b**) nondimensional velocity v* versus time.

4.2.4. Effect of the Shear Coefficient of the Stub

Finally, we conduct a study on the effect of the shear coefficient of the stub on the non-dimensional deflection and velocity of the centroid point of the shell. We consider four cases with $k_s = 10^3$, 10^5, 10^{10}, 10^{12}, and 10^{15} Pa. Geometrical parameters are $a = b$; $h = a/50$, $h_c/h_t = 2$, $R/a = 6$. The numerical results of non-dimensional deflection and velocity of the centroid point of the shell are plotted in Figure 8, the maximum value is shown in Table 9.

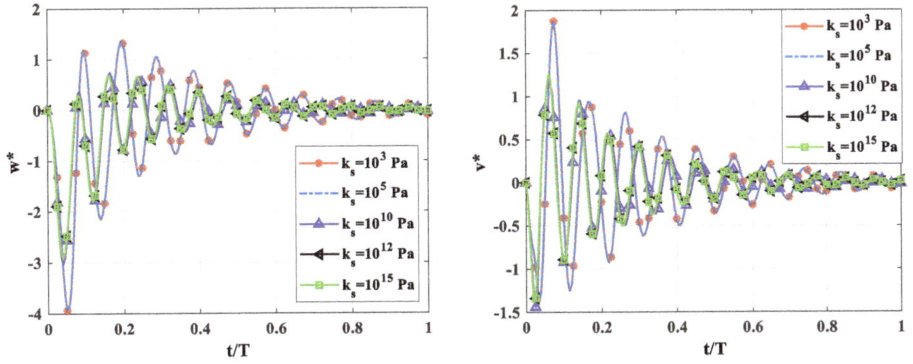

(a) (b)

Figure 8. Effect of shear coefficient of stub on the non-dimensional deflection and velocity of the centroid point. (**a**) Nondimensional deflection w* versus time; and (**b**) nondimensional velocity v* versus time.

Table 9. Effect of shear coefficient of stub on the non-dimensional deflection and velocity of the centroid point.

Maximum Values	k_s=10^3Pa	k_s=10^5Pa	k_s=10^{10}Pa	k_s=10^{12}Pa	k_s=10^{15}Pa
w^*_{max}	3.9352	3.9352	3.0341	2.9070	2.9059
v^*_{max}	1.3658	1.3658	1.4354	1.3657	1.3662

In this last case, we can see in Figure 8 and Table 9 that when the shear coefficient of the stub increases, the non-dimensional deflection and velocity of the centroid point of the shell is reduced. It is easily understood that the enhancing of the stiffness of the stud makes the total structure get stronger, meaning the stiffness of the shell is increased, correspondingly.

4.2.5. Influence of the Mass Density of the Core Layer

Let us consider a four-edge simply supported (SSSS) shell ($b = a$) with h_c = h/2, $h_t = h_b$ = h/4. The shear modulus of the shear connector is k_s = 50 MPa. The mass densities of the three layers are $\rho_t = \rho_b = 2300$ kg/m^3 and $\rho_c = 700,\ 1000,\ 1500,\ 2000,\ 2300$ kg/m^3. Nondimensional deflection and velocity of the shell center point are shown in Figure 9, maximum deflections and velocities of the shell center point are illustrated in Table 10. The mass ratio of the shell corresponding to the different values of ρ_c compared to case $\rho_t = \rho_c = \rho_b = 2300$ kg/m^3 is shown in Table 11.

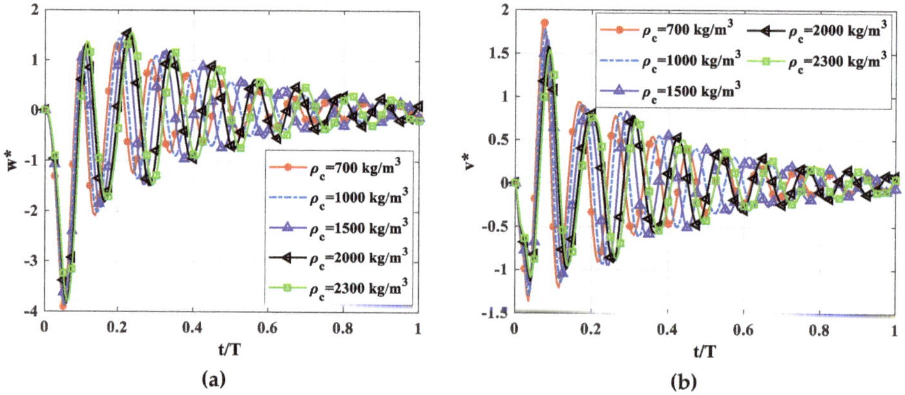

Figure 9. Dynamic deflections of center point of the plate versus time for different ρ_c. (**a**) Nondimensional deflection w* versus time, and (**b**) nondimensional velocity v* versus time.

Table 10. Maximum deflections, velocities and stress of the shell center point versus time for different ρ_c.

Maximum Values	ρ_c= 700 (kg/m^3)	ρ_c= 1000 (kg/m^3)	ρ_c= 1500 (kg/m^3)	ρ_c= 2000 (kg/m^3)	ρ_c= 2300 (kg/m^3)
w^*_{max}	3.9020	3.8837	3.8403	3.8318	3.7942
v^*_{max}	1.3653	1.2962	1.1985	1.1339	1.0871

Table 11. The mass ratio of the shell corresponding to the different values of ρ_c.

Mass Density of the Core Layer	ρ_c= 700 (kg/m^3)	ρ_c= 1000 (kg/m^3)	ρ_c= 1500 (kg/m^3)	ρ_c= 2000 (kg/m^3)	ρ_c= 2300 (kg/m^3)
The mass ratio ($100\frac{\rho_c+\rho_t}{2\rho_t}$%)	65.21	71.73	82.60	93.44	100
Reduced mass (%)	34.79	28.27	17.40	6.56	0

Comment: From the Figure 9 and Tables 10 and 11, we obtain that when the mass density of the core-layer is increased from 700 to 2300 kg, deflection and velocity of the shell center point are almost not changed. Therefore, in order to reduce the mass of the shell, we can use the triple-layer shell with shear connectors, which the core layer has a smaller mass density than other layers. Specifically, corresponding to a difference of mass density of the core layer ρ_c = 700, 1000, 1500, 2000 kg/m^3, the mass of the shell decreases by 34.79, 28.27, 17.40, and 6.56%.

4.2.6. Influence of Modulus of Elasticity

Let us consider a fully simply-supported (SSSS) shell ($b = a$) with h_c = h/2, $h_t = h_b$ = h/4. The shear modulus of the shear connector is k_s = 50 MPa. The modulus of elasticity of the three layers are $E_t = E_b$ = 12 GPa and E_c = 8, 9, 10, 12 GPa. Nondimensional deflection and velocity of the shell center point are shown in Figure 10, and the maximum deflections and velocities of the shell center point are illustrated in Table 12.

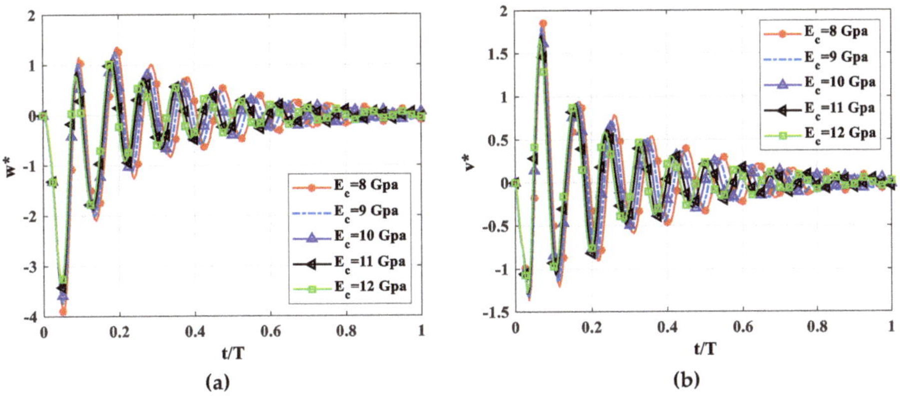

Figure 10. Dynamic deflections of center point of the shell over time with different E_c. (**a**): Nondimensional deflection w* versus time, and (**b**) nondimensional velocity v* versus time.

Table 12. Maximum deflection and velocity of the shell center point over time for different E_c.

Maximum Values	E_c = 8 GPa	E_c = 9 GPa	E_c = 10 GPa	E_c = 12 GPa	E_c = 12 GPa
w^*_{max}	3.9020	3.7494	3.5895	3.4252	3.3031
v^*_{max}	1.3653	1.3375	1.2998	1.2841	1.2792

Comment: From the Figure 10 and Table 12, we can find that when modulus of elasticity of the core-layer is increased in a range from 8 to 12 GPa, deflection and velocity of the shell center point are slightly decreased.

5. Conclusions

The finite element method (FEM) is the numerical method for solving problems of engineering and mathematical physics, including the calculation of shell structures. Establishing the balance equation describing the vibration of shell structure is quite simple and it is very convenient for coding on a personal computer (PC). The proposed program is able to analyze the static bending, dynamic response, nonlinear problems, etc., with complicated structures, which are not easy to solve by analytical methods.

Based on the finite element method, we established the equilibrium equation of a triple-layer composite shell with shear connectors subjected to dynamic loads. In this paper, employing of the eight-node isometric element is suitable. To exactly describe the strain field, the displacements of the three-layer shell with shear connectors, and the 13-degrees of freedom element is used, in which the three layers have the same a degree of freedom in the z-direction, and the other 12 degrees of freedom are described as the linear displacement and rotation angle of each layer. Hence, the displacement field and the strain field of each layer can be investigated deeply. We have created the program in the MATLAB environment to investigate effects of various geometrical parameters on free and forced vibrations of shells. To sum up, some main interesting points of this paper are listed in the following statements.

In general, the geometrical parameters effect strongly on free and forced vibrations of the shell; when the shape of the shell is small, the structure gets stiffer.

Based on the numerical results, we realized that for this type of structure, the shear coefficient of the stub plays a very important role. Especially, when the stiffness of the shear coefficient is large enough, this structure seems to be a sandwich shell.

From the above computed results, we suggest that in order to reduce the vibration of such a structure, we should use the middle layer, having the elastic modulus less than other layers, and the thickness of the middle layer 20–30 times larger than the other ones.

We suggest that, in order to reduce the volume of the shell structure subjected to the blast load, we should consider the triple-layer shell with the core layer having a smaller density than the two layers others. Another interesting thing is that the core layer has less stiffener than the other two layers for the displacement response, the velocity is almost unchanged, so we can be flexible in making shells with available materials and different stiffeners.

Based on the achieved numerical results, this paper also leads to further works; for instance, the analysis of FGM structures with shear connectors, buckling problems, the composite plate with shear connectors subjected to both temperature and mechanical loads, and so on.

Author Contributions: Investigation, H.-N.N.; Software, T.N.C., T.T.T.; Visualization, T.V.K.; Writing–original draft, V.-D.P.; Writing–review & editing, D.V.T.

Funding: This research was funded by Vietnam National Foundation for Science and Technology Development (NAFOSTED) grant number 107.02-2018.30.

Acknowledgments: DVT gratefully acknowledges the supports of Vietnam National Foundation for Science and Technology Development (NAFOSTED) under grant number 107.02-2018.30.

Conflicts of Interest: The authors declare no conflict of interest.

Appendix A

$$\mathbf{B}_{ti}^{0} = \begin{bmatrix} \frac{\partial N_i}{\partial x} & 0 & 0 & 0 & 0 & 0 & 0 & 0 & 0000\frac{N_i}{R_x} \\ 0 & \frac{\partial N_i}{\partial y} & 0 & 0 & 0 & 0 & 0 & 0 & 0000\frac{N_i}{R_y} \\ \frac{\partial N_i}{\partial y} & \frac{\partial N_i}{\partial x} & 0 & 0 & 0 & 0 & 0 & 0 & 0000\frac{2N_i}{R_{xy}} \end{bmatrix}$$

$$\mathbf{B}_{ci}^{0} = \begin{bmatrix} 0 & 0 & 0 & 0 & \frac{\partial N_i}{\partial x} & 0 & 0 & 0 & 0000\frac{N_i}{R_x} \\ 0 & 0 & 0 & 0 & 0 & \frac{\partial N_i}{\partial y} & 0 & 0 & 0000\frac{N_i}{R_y} \\ 0 & 0 & 0 & 0 & \frac{\partial N_i}{\partial y} & \frac{\partial N_i}{\partial x} & 0 & 0 & 0000\frac{2N_i}{R_{xy}} \end{bmatrix}$$

$$\mathbf{B}_{bi}^{0} = \begin{bmatrix} 0 & 0 & 0 & 00000 & \frac{\partial N_i}{\partial x} & 0 & 0 & 0 & \frac{N_i}{R_x} \\ 0 & 0 & 0 & 00000 & 0 & \frac{\partial N_i}{\partial y} & 0 & 0 & \frac{N_i}{R_y} \\ 0 & 0 & 0 & 00000 & \frac{\partial N_i}{\partial y} & \frac{\partial N_i}{\partial x} & 0 & 0 & \frac{2N_i}{R_{xy}} \end{bmatrix}$$

$$\mathbf{B}_{ti}^{1} = \begin{bmatrix} 0 & 0 & \frac{\partial N_i}{\partial x} & 0 & 0 & 0 & 0 & 0 & 00000 \\ 0 & 0 & 0 & \frac{\partial N_i}{\partial y} & 0 & 0 & 0 & 0 & 00000 \\ -C_0\frac{\partial N_i}{\partial y} & C_0\frac{\partial N_i}{\partial x} & \frac{\partial N_i}{\partial y} & \frac{\partial N_i}{\partial x} & 0 & 0 & 0 & 0 & 00000 \end{bmatrix}$$

$$\mathbf{B}_{ci}^{1} = \begin{bmatrix} 0 & 0 & 0 & 0 & 0 & 0 & \frac{\partial N_i}{\partial x} & 0 & 00000 \\ 0 & 0 & 0 & 0 & 0 & 0 & 0 & \frac{\partial N_i}{\partial y} & 00000 \\ 0 & 0 & 0 & 0 & -C_0\frac{\partial N_i}{\partial y} & C_0\frac{\partial N_i}{\partial x} & \frac{\partial N_i}{\partial y} & \frac{\partial N_i}{\partial x} & 00000 \end{bmatrix}$$

$$\mathbf{B}_{bi}^{1} = \begin{bmatrix} 0 & 0 & 0 & 00000 & 0 & 0 & \frac{\partial N_i}{\partial x} & 0 & 0 \\ 0 & 0 & 0 & 00000 & 0 & 0 & 0 & \frac{\partial N_i}{\partial y} & 0 \\ 0 & 0 & 0 & 00000 & -C_0\frac{\partial N_i}{\partial y} & C_0\frac{\partial N_i}{\partial x} & \frac{\partial N_i}{\partial y} & \frac{\partial N_i}{\partial x} & 0 \end{bmatrix}$$

$$\mathbf{S}_{ti} = \begin{bmatrix} -\frac{N_i}{R_x} & -\frac{N_i}{R_{xy}} & 0 & N_i & 00000 & 0 & 0 & 0 & \frac{\partial N_i}{\partial y} \\ -\frac{N_i}{R_{xy}} & -\frac{N_i}{R_y} & N_i & 0 & 00000 & 0 & 0 & 0 & \frac{\partial N_i}{\partial x} \end{bmatrix}$$

$$\mathbf{S}_{ci} = \begin{bmatrix} 0 & 0 & 0 & 0 & -\frac{N_i}{R_x} & -\frac{N_i}{R_{xy}} & 0 & N_i0000 & \frac{\partial N_i}{\partial y} \\ 0 & 0 & 0 & 0 & -\frac{N_i}{R_{xy}} & -\frac{N_i}{R_y} & N_i & 00000 & \frac{\partial N_i}{\partial x} \end{bmatrix}$$

$$\mathbf{S}_{bi} = \begin{bmatrix} 0 & 0 & 0 & 00000 & -\frac{N_i}{R_x} & -\frac{N_i}{R_{xy}} & 0 & N_i & \frac{\partial N_i}{\partial y} \\ 0 & 0 & 0 & 00000 & -\frac{N_i}{R_{xy}} & -\frac{N_i}{R_y} & N_i & 0 & \frac{\partial N_i}{\partial x} \end{bmatrix}$$

in which $C_0 = \frac{1}{R_y} - \frac{1}{R_x}$.

Appendix B

$$\mathbf{L}_t = \begin{bmatrix} 10z_t0000000000 \\ 010z_t000000000 \\ 0000000000001 \end{bmatrix}$$
$$\mathbf{L}_c = \begin{bmatrix} 000010z_c000000 \\ 0000010z_c00000 \\ 0000000000001 \end{bmatrix}$$
$$\mathbf{L}_b = \begin{bmatrix} 0000000010z_b00 \\ 00000000010z_b0 \\ 0000000000001 \end{bmatrix}$$

References

1. Newmark, N.M.; Siess, C.P.; Viest, I.M. Test and analysis of composite beams with incomplete interaction. *Proc. Soc. Exp. Stress Anal.* **1951**, *19*, 75–92.
2. He, G.; Yang, X. Finite element analysis for buckling of two-layer composite beams using Reddy's higher order beam theory. *Finite Elem. Anal. Des.* **2014**, *83*, 49–57. [CrossRef]
3. Xu, R.; Wang, G. Variational principle of partial-interaction composite beams using Timoshenko's beam theory. *Int. J. Mech. Sci.* **2012**, *60*, 72–83. [CrossRef]
4. Xu, R.; Wu, Y. Static, dynamic, and buckling analysis of partial interaction composite members using Timoshenko's beam theory. *Int. J. Mech. Sci.* **2007**, *49*, 1139–1155. [CrossRef]
5. Nguyen, Q.H.; Hjiaj, M.; Grognec, P.L. Analytical approach for free vibration analysis of two-layer Timoshenko beams with interlayer slip. *J. Sound Vib.* **2012**, *331*, 2949–2961. [CrossRef]
6. Da Silva, A.R.; Sousa, J.B.M., Jr. A family of interface elements for the analysis of composite beams with interlayer slip. *Finite Elem. Anal. Des.* **2009**, *45*, 305–314. [CrossRef]
7. Schnabl, S.; Saje, M.; Turk, G.; Planinc, I. Analytical solution of two-layer beam taking into account interlayer slip and shear deformation. *J. Struct. Eng.* **2007**, *133*, 886–894. [CrossRef]
8. Nguyen, Q.H.; Martinelli, E.; Hjiaj, M. Derivation of the exact stiffness matrix for a two-layer Timoshenko beam element with partial interaction. *Eng. Struct.* **2011**, *33*, 298–307. [CrossRef]
9. Nguyen, Q.H.; Hjiaj, M.; Lai, V.-A. Force-based FE for large displacement inelastic analysis of two-layer Timoshenko beams with interlayer slips. *Finite Elem. Anal. Des.* **2014**, *85*, 1–10. [CrossRef]
10. Huang, C.W.; Su, Y.H. Dynamic characteristics of partial composite beams. *Int. J. Struct. Stab. Dyn.* **2008**, *8*, 665–685. [CrossRef]
11. Shen, X.; Chen, W.; Wu, Y.; Xu, R. Dynamic analysis of partial-interaction composite beams. *Compos. Sci. Technol.* **2011**, *71*, 1286–1294. [CrossRef]

12. Arvin, H.; Bakhtiari-Nejad, F. Nonlinear free vibration analysis of rotating composite Timoshenko beams. *Compos. Struct.* **2013**, *96*, 29–43. [CrossRef]
13. Chakrabarti, A.; Sheikh, A.H.; Griffith, M.; Oehlers, D.J. Dynamic response of composite beams with partial shear interaction using a higher-order beam theory. *J. Struct. Eng.* **2013**, *139*, 47–56. [CrossRef]
14. Chakrabarti, A.; Sheikh, A.H.; Griffith, M.; Oehlers, D.J. Analysis of composite beams with longitudinal and transverse partial interactions using higher order beam theory. *Int. J. Mech. Sci.* **2012**, *59*, 115–125. [CrossRef]
15. Chakrabarti, A.; Sheikh, A.H.; Griffith, M.; Oehlers, D.J. Analysis of composite beams with partial shear interactions using a higher order beam theory. *Eng. Struct.* **2012**, *36*, 283–291. [CrossRef]
16. Subramanian, P. Dynamic analysis of laminated composite beams using higher order theories and finite elements. *Compos. Struct.* **2006**, *73*, 342–353. [CrossRef]
17. Li, J.; Shi, C.; Kong, X.; Li, X.; Wu, W. Free vibration of axially loaded composite beams with general boundary conditions using hyperbolic shear deformation theory. *Compos. Struct.* **2013**, *97*, 1–14. [CrossRef]
18. Vo, T.P.; Thai, H.-T. Static behavior of composite beams using various refined shear deformation theories. *Compos. Struct.* **2012**, *94*, 2513–2522. [CrossRef]
19. Tarun, K.; Owen, D.R.J.; Zienkiewicz, O.C. A refined higher-order C^0 plate bending element. *Compos. Struct.* **1982**, *15*, 177–183.
20. Manjunatha, B.S.; Kant, T. New theories for symmetric/unsymmetric composite and sandwich beams with C0 finite elements. *Compos. Struct.* **1993**, *23*, 61–73. [CrossRef]
21. Yan, J.B.; Zhang, W. Numerical analysis on steel-concrete-steel sandwich plates by damage plasticity model, Materials to structures. *Constr. Build. Mater.* **2017**, *149*, 801–815. [CrossRef]
22. Carrera, E.; Petrolo, M.; Zappino, E. Performance of CUF Approach to Analyze the Structural Behavior of Slender Bodies. *J. Struct. Eng.* **2012**, *138*, 285–297. [CrossRef]
23. Carrera, E. Theories and finite elements for multilayered, anisotropic, composite plates and shells. *Arch. Comput. Methods Eng.* **2002**, *9*, 87–140. [CrossRef]
24. Cinefra, M.; Kumar, S.K.; Carrera, E. MITC9 Shell elements based on RMVT and CUF for the analysis of laminated composite plates and shells. *Compos. Struct.* **2019**, *209*, 383–390. [CrossRef]
25. Muresan, A.A.; Nedelcu, M.; Gonçalves, R. GBT-based FE formulation to analyse the buckling behaviour of isotropic conical shells with circular cross-section. *Thin-Walled Struct.* **2019**, *134*, 84–101. [CrossRef]
26. Schardt, R. *Verallgemeinerte Technische Biegetheorie: Lineare Probleme*; Springer: Berlin/Heidelberg, Germany, 1989.
27. Fryba, L. *Vibration of Solids and Structures under Moving Loads*; Springer: Berlin, Germany, 1999.
28. Hoang-Nam, N.; Tan-Y, N.; Ke, V.T.; Thanh, T.T.; Truong-Thinh, N.; Van-Duc, P.; Thom, V.D. A Finite Element Model for Dynamic Analysis of Triple-Layer Composite Plates with Layers Connected by Shear Connectors Subjected to Moving Load. *Materials* **2019**, *12*, 598–617.
29. Ansari Sadrabadi, S.; Rahimi, G.H.; Citarella, R.; Shahbazi Karami, J.; Sepe, R.; Esposito, R. Analytical solutions for yield onset achievement in FGM thick walled cylindrical tubes undergoing thermomechanical loads. *Compos. Part B Eng.* **2017**, *116*, 211–223. [CrossRef]
30. Bathe, K.J. *Finite element Procedures*; Prentice-Hall International Inc.: Upper Saddle River, NJ, USA, 1996.
31. Wolf, J.P. *Dynamic Soil-Structure Interaction*; Prentice-Hall Inc.: Upper Saddle River, NJ, USA, 1985.
32. Reddy, J.N. *Mechanics of Laminated Composite Plate and Shell*, 2nd ed.; CRC Press: Boca Raton, FL, USA, 2004.
33. Qian, L.F. Free and Forced Vibrations of Thick Rectangular Plates using Higher-Order Sheara and Normal Deformable Plate Theory and Meshless Petrov-Galerkin (MLPG) Method. *CMES* **2003**, *5*, 519–534.

symmetry

MDPI

Article

Towards Infinite Tilings with Symmetric Boundaries

Florian Stenger [1] **and Axel Voigt** [1,2,*]

[1] Institute of Scientific Computing, TU Dresden, 01062 Dresden, Germany; florian.stenger@tu-dresden.de
[2] Dresden Center for Computational Materials Science (DCMS), TU Dresden, 01062 Dresden, Germany
* Correspondence: axel.voigt@tu-dresden.de; Tel.: +49-351-46334187

Received: 11 February 2019; Accepted: 22 March 2019; Published: 27 March 2019

Abstract: Large-time coarsening and the associated scaling and statistically self-similar properties are used to construct infinite tilings. This is realized using a Cahn–Hilliard equation and special boundaries on each tile. Within a compromise between computational effort and the goal to reduce recurrences, an infinite tiling has been created and software which zooms in and out evolve forward and backward in time as well as traverse the infinite tiling horizontally and vertically. We also analyze the scaling behavior and the statistically self-similar properties and describe the numerical approach, which is based on finite elements and an energy-stable time discretization.

Keywords: symmetric boundary condition; pattern formation; computational design; finite-element method

1. Introduction

At first-order phase transitions, coexisting macroscopic domains of different phases emerge from small fluctuations of a homogeneous phase. Late stages of this process are often dominated by the motion of the interfaces separating the domains. Considering this large-time coarsening behavior, i.e., the growth of a characteristic length scale $l(t)$ as $t \to \infty$ determines important characteristics of the dynamics and led to the identification of several universality classes of domain growth. We are here concerned with conserved order parameters, for which the expected behavior is $l(t) \sim t^{1/3}$, which results from the scale invariance of the Mullins–Sekerka system $x \to \lambda x$, $t \to \lambda^3 t$. Rigorous results exist for an upper bound for $l(t)$, stating that microstructures cannot coarsen faster than the similarity rate [1]. As there are non-generic configurations, e.g., stripe domains with zero curvature which are stable, lower bounds cannot be expected within a deterministic framework. Besides this scaling law, solutions with random initial data are also believed to be statistically self-similar in this large-time regime. Numerical studies based on a Cahn–Hilliard equation and related coarse-grained theories indicate that the approach of the large-time regime with the statistically self-similar structures might be very slow [2]. To explore these regimes numerically thus requires large length and time scales, which limits the accessible sample size. We are here interested in these statistically self-similar structures, which have been used for various art and design projects, e.g., [3]. Here, we would like to explore very large, in principle infinite, samples. To tackle such a system we consider, instead of one huge simulation, many moderately sized domains with different initial data, and require the boundaries to match. If appropriately done, this will allow construction of large (infinite) tilings which are statistically self-similar. With a random arrangement of finitely many computed structures, the impression of an infinite tiling with no recurrence could be achieved. For this impression, the boundary conditions at the computational domains are crucial. They are described in detail in Section 4 together with the finite-element approach to solve the Cahn–Hilliard equation. In Section 2 we show various results, among other things a computer program which allows navigation through space and time of an infinitely extended structure. We further discuss improvements and outline possible applications. In Section 3, we discuss scaling and self-similar properties.

2. Results

As the underlying model for phase transitions with a conserved order parameter we consider a Cahn–Hilliard equation

$$\partial_t \phi = \gamma \Delta \mu, \qquad \mu = -\epsilon^2 \Delta \phi + B'(\phi); \tag{1}$$

see Section 4 for details. First, we consider coarsening in a rectangular cuboid using standard boundary conditions $\mathbf{n} \cdot \nabla \phi = \mathbf{n} \cdot \nabla \mu = 0$ on $\partial \Omega$. Figure 1 shows snapshots of the results within the large-time regime, visualized in various ways. The interface area is minimized, and the structure thus coarsens in time. To analyze this in a statistical manner requires either a larger domain or more samples. Our idea is to combine both by using samples which are distinct from each other but fit together to form a large and extendable structure. The boundary conditions in the current setting enforce the level lines of the interface to be perpendicular to $\partial \Omega$. They in addition do not fit to each other and thus do not allow combination of different samples. To overcome these limitations, we consider a smaller domain, again a rectangular cuboid and the boundary conditions introduced in Section 4 which specify the values of ϕ and the normal flux $\nabla \phi \cdot \mathbf{n}$ such that opposite sides match. The first approach only has two distinct boundaries, $N = 2$. Figure 2 shows four samples, all obtained with different initial conditions and considered at the same time instance. The structures fit together and any translation in x- or y-direction by the width of the domain will also fit. The individual figures are provided in SI as Figures S1–S4; print them and try it out. The figures are part of an art project, $M = 100$ individual samples have been computed and printed on Alu-Dibond in size 20 cm $\times$ 20 cm, creating a 200 cm $\times$ 200 cm figure which can be displayed in $100! \approx 9.332622 \cdot 10^{157}$ variations.

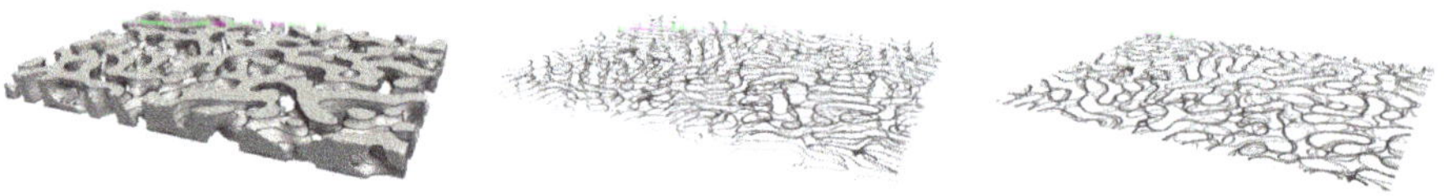

Figure 1. Typical structure within the large-time regime, visualized as $\phi = 1$ in Ω, $\phi = 0.5$ at $z = 0.01, 0.13, 0.25, 0.37, 0.49$ and $\phi = 0.5$ at $z = 0.01, 0.13, 0.25, 0.37, 0.49$ projected to $z = 0$, from left to right. The boundary conditions are $\mathbf{n} \cdot \nabla \phi = \mathbf{n} \cdot \nabla \mu = 0$ on $\partial \Omega$. The corresponding videos of the coarsening process are provided in SI as Videos S5–S7.

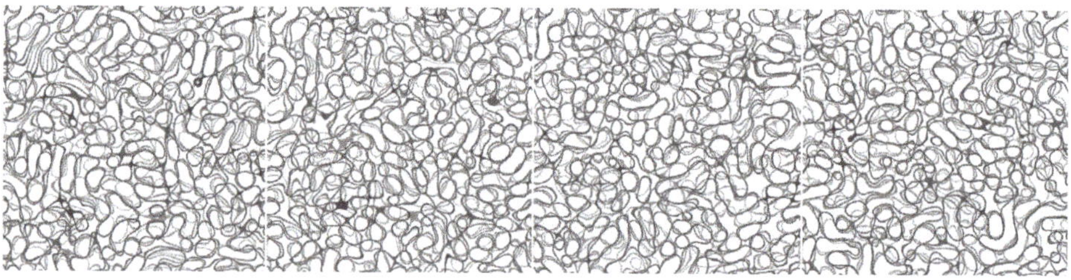

Figure 2. Four samples with identical boundaries but distinct inner structure. The samples are translational invariant in x- and y-direction.

Even if the inner structure is unique for each tile, the boundaries in x- and y-direction are the same and recurrences are visible. To improve on this issue, we consider a second approach, which is less flexible in terms of arrangement of samples but minimizes possible recurrences. As a compromise of computational cost and visual impression we consider tilings with $N = 10$ different boundary conditions. This improves the impression of a tiling with no recurrence since systematic recurrence in a row, a column, or diagonally can be avoided by careful assembly of the tiles. Repetitions do not only appear less frequently but also in a pattern that is much less obvious. To see this recurrence without knowing the construction process and explicitly searching for them is almost impossible. We consider two different internal realizations each, leading to $M = 10$. Within the proposed pattern determined by the boundary conditions the inner realization are randomly chosen to construct an infinite tiling, where recurrences are almost invisible.

A software is developed to visualize the infinite structure. We consider visualizations with five projected level lines of the interface. The software allows zooming in and out, evolve forward and backward in time as well as traverse the infinite tiling horizontally and vertically. Figure 3 shows some screenshots, starting from an early time instant and a low zoom factor (a), going to a late-time instant of this setting (b), zooming into the structure (c), evolving along a trajectory in space and time, which keeps the interface area constant (d), and going back to the initial state (a). Videos of the journey through space and time are provided in SI as Videos S8 and S9.

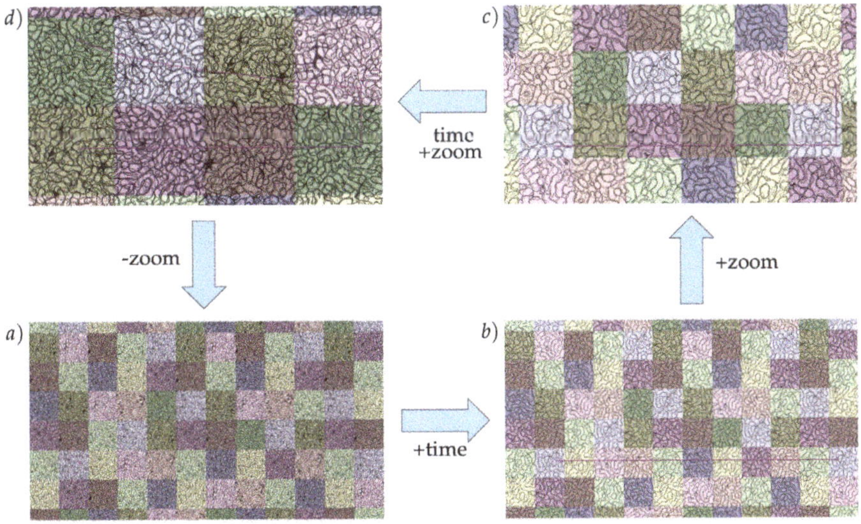

Figure 3. Screenshots of the visualization software, here in addition color-coded according to the individual tiles used. The dark magenta lines indicate the user-interaction. Moving the mouse horizontally evolves time, moving it vertically zooms in and out.

As the proposed approach is in principle not restricted to rectangular cuboidal domains various possibilities for applications can be imagined. Besides wallpaper design, they range from fashion design with individualized clothes to camouflage patterns of automotive prototypes. Here, we highlight a more entertaining application, a Rubik's cube which always fits, but has 24 different fields; see Figure 4.

Figure 4. A Rubik's cube which always fits even if all tiles are different. A video is provided in SI as Video S10.

3. Discussion

We now explore the scaling properties and the statistically self-similar behavior of the constructed infinite tiling. The theoretically scaling behavior $l(t) \sim t^{1/3}$ of a characteristic length scale is tested by computing the interface area of each tile over time. The interface is thereby represented as the level-surface $\phi = 0$. In addition, we also compute the length of the interface of the level line at $z = 0.25$, which is used for visualization. Figure 5 shows the results over time, which are averaged over all individual realizations of tiles.

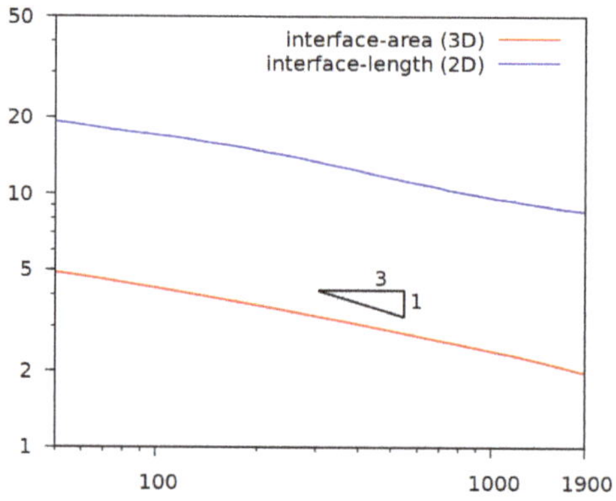

Figure 5. Development of the interface area of the 3D-structure and the interface-length of a slice through the center of the domain over time, averaged for all tiles.

The results lead to a scaling exponent which is below the upper bound of 1/3. The slope is not constant, but on average equal for the two measures and approximately 0.26. There are different reasons for this lower value, either the structure has not reached the late-time behavior for which the theoretical scaling behavior is expected, or the considered domain with the zero-flux boundary conditions on top and bottom favor parallel structures and thus prevent coarsening. The limitation of our approach, to combine several tiles and to compute them separately, should also be mentioned. This approach only allows the consideration of coarsening up to a length scale of the size of a tile. For larger times, the approach is no longer valid. However, even if the theoretical scaling law could not be shown computationally, statistical self-similarity still might be possible. Statistical self-similarity can already be expected from a randomly chosen sample. We consider the middle slice at an early time instant, its coarsening and subsequent zooming-out in Figure 6. Instead of the interface line the two phases are rendered in black and white. When the coarse structure is zoomed out to a degree where the interface-length matches that of the earlier timestep, both structures are visually similar (see Figure 6a,c). To quantify this, we compute for each row and column of pixels the distance between two interfaces. This is done for all samples and plotted for different times in Figure 7.

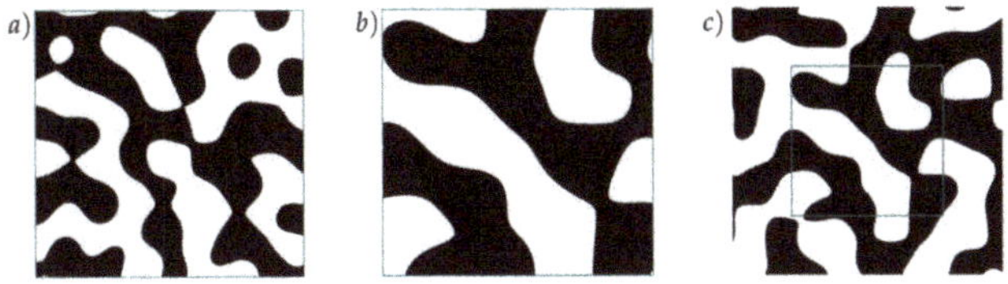

Figure 6. Middle slice of the domain in an early time instance (**a**), a later point in time (**b**) and the later time-point zoomed out until the interface-length equals that of the early one (**c**). In the last image the unzoomed region is framed to indicate the level of zoom applied.

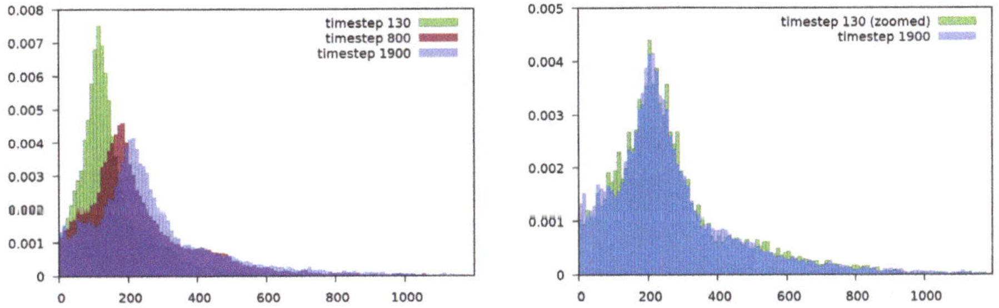

Figure 7. Left: density-histograms of the distances between interfaces for three selected timesteps (all tiles combined). **Right**: a square subregion of timestep 130 is considered which, after zooming to the size of a full tile, has the same interface area as timestep 1900. The histograms match almost exactly, which computationally indicates the statistically self-similar structure.

Even if the theoretically predicted scaling law $l(t) \sim t^{1/3}$ could not be computationally shown, the large-scale simulations, which run for each tile on a high-performance computer, reproduce the predicted statistical self-similarity. The huge structure, which results as an arrangement of individual tiles, would not have been possible to simulate on the available hardware. The approach fulfills two goals, it provides enough statistics to analyze scaling and statistical self-similarity and it allows the generation of aesthetically appealing tilings with almost invisible recurrences, which can be infinitely extended.

4. Materials and Methods

The Cahn–Hilliard equation [4] is a fourth order partial differential equation resulting as a H^{-1} gradient flow of a Ginzburg-Landau energy

$$\mathcal{E}[\phi] = \int_\Omega \gamma \left(\frac{\epsilon^2}{2} |\nabla \phi|^2 + B(\phi) \right) \, d\Omega, \tag{2}$$

where Ω is a bounded domain, ϵ a positive parameter determining the width of the diffuse interface, γ the surface energy, here considered as a positive constant, and $B(\phi) = \frac{1}{4}(1 - \phi^2)^2$ a double well potential. The resulting equation reads

$$\partial_t \phi = \gamma \Delta \mu, \qquad \mu = -\epsilon^2 \Delta \phi + B'(\phi) \tag{3}$$

and converges for $\epsilon \to 0$ to the Mullins–Sekerka problem [5]; see Ref. [6].

Various numerical approaches have been proposed to solve the equation efficiently. We consider a convexity splitting approach, e.g., [7–11]. The idea is to split the double well potential $B(\phi) = B_c(\phi) - B_e(\phi)$, such that both parts are convex and to consider the time discretization as

$$d_\tau \phi^{n+1} = \gamma \Delta \mu^{n+1}, \qquad \mu^{n+1} = -\epsilon^2 \Delta \phi^{n+1} + B_c'(\phi^{n+1}) - B_e'(\phi^n), \tag{4}$$

with discrete time derivative $d_\tau \phi^{n+1} = (\phi^{n+1} - \phi^n)/\tau_n$. The resulting scheme is unconditionally energy stable, unconditionally solvable and converges optimally in the energy norm [9]. To solve the above systems, we consider a linearization of $B_c'(\phi^{n+1}) \approx B_c'(\phi^n) + B_c''(\phi^n)(\phi^{n+1} - \phi^n)$. We further consider adaptive mesh refinement according to criteria related to the position of the diffuse interface, here $\nabla\phi$, to ensure a resolution of approximately five grid points across the interface and a coarser mesh elsewhere; see Figure 8.

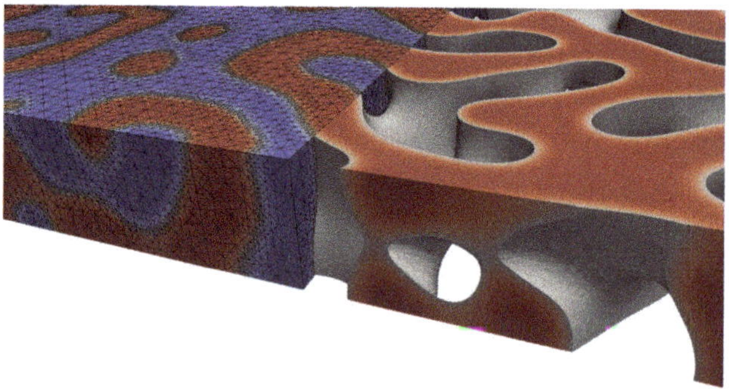

Figure 8. Typical structure, highlighting one of the two phases and the adaptively refined mesh along the diffuse interface.

The resulting linear system is solved in parallel using a block-preconditioner, see [12,13], and the iterative solver FGMRES. All problems are implemented in the adaptive finite-element toolbox AMDiS [14,15]. The considered parameters are $\epsilon = 0.01$ and $\gamma = 1.0$ and as computational domain rectangular cuboid $\Omega = (0, L)x(0, L)x(0, l)$ with $l = 0.2$ and $L = 2.0$ for the larger and $L = 1.0$ for the smaller domain and boundaries Γ^{top}, Γ^{bottom}, Γ^0 and $\Gamma^1 = \bigcup_{i=1}^4 \Gamma^{1,i}$, see Figure 9. The number of grid points on the larger domain reduces from approx. 2.95 million at the beginning to approx. 1.95 million at the final configuration.

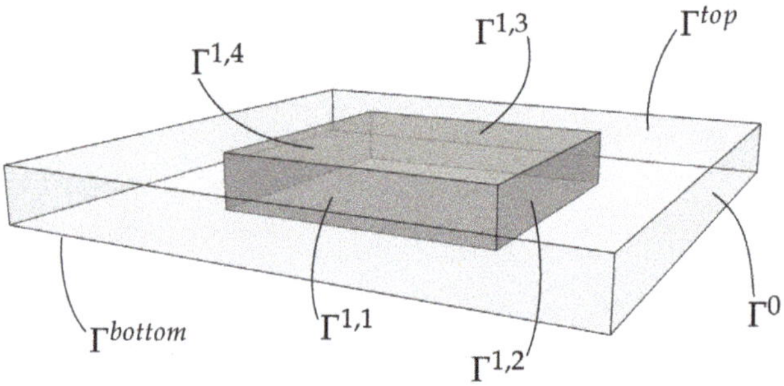

Figure 9. Geometric setting and boundaries.

As initial condition we consider white noise around the mean value $\phi = 0$. At Γ^{top} and Γ^{bottom} we specify zero-flux boundary conditions for ϕ and μ. We first consider the larger domain with zero-flux boundary conditions for ϕ and μ also on Γ^0. In this setting the finite-element formulation in each time step reads: Find $\phi^{n+1}, \mu^{n+1} \in V_h$ such that $\forall \eta, \xi \in V_h$

$$\int_\Omega d_\tau \phi^{n+1} \eta \, dx + \gamma \int_\Omega \nabla \mu^{n+1} \cdot \nabla \eta \, dx = 0 \tag{5}$$

$$\int_\Omega \mu^{n+1} \xi \, dx - \epsilon^2 \int_\Omega \nabla \phi^{n+1} \cdot \nabla \xi \, dx - \int_\Omega B_c'(\phi^{n+1}) \xi \, dx = -\int_\Omega B_e'(\phi^n) \xi \, dx, \tag{6}$$

with $V_h = \{v \in C^0(\Omega) | v_{|T} \in P^1(T) \forall T \in \mathcal{T}\}$ the space of piecewise linear Lagrange elements and triangulation $\mathcal{T}$. In each time step we extract $\phi_1^{n+1} = \phi^{n+1}$ and $\mathbf{n} \cdot \nabla \phi_1^{n+1} = \mathbf{n} \cdot \nabla \phi^{n+1}$ along the inner boundary Γ^1. These data are going to be used in the subsequent computations in the smaller domain as boundary conditions on Γ^1. The finite-element formulation now reads in each time step: Find $\phi^{n+1} \in V_{h,\Gamma^1,\phi_1}$ and $\mu^{n+1} \in V_h$ such that $\forall \eta \in V_{h,\Gamma^1,0}$ and $\forall \xi \in V_h$

$$\int_\Omega d_\tau \phi^{n+1} \eta \, dx + \gamma \int_\Omega \nabla \mu^{n+1} \cdot \nabla \eta \, dx = 0 \tag{7}$$

$$\int_\Omega \mu^{n+1} \xi \, dx - \epsilon^2 \int_\Omega \nabla \phi^{n+1} \cdot \nabla \xi \, dx - \int_\Omega B_c'(\phi^{n+1}) \xi \, dx = -\int_\Omega B_e'(\phi^n) \xi \, dx - \epsilon^2 \int_{\Gamma^1} \mathbf{n} \cdot \nabla \phi_1 \xi ds, \tag{8}$$

with $V_{h,\Gamma^1,\alpha} = \{v \in C^0(\Omega) | v_{|T} \in P^1(T) \forall T \in \mathcal{T}, v = \alpha \text{ on } \Gamma^1\}$.

For different initial data this leads to different solutions with common boundary conditions. However, to construct tilings, they also must match, which is not yet guaranteed. To fulfill this requirement, we proceed in two different ways. The first approach considers only one computation on the larger domain and uses the extracted values and fluxes ϕ_1^{n+1} and $\mathbf{n} \cdot \nabla \phi_1^{n+1}$ at Γ^1 only from two sides $\Gamma^{1,1}$ and $\Gamma^{1,2}$ ($N = 2$) and specifies them also on the opposite sides for the computations on the smaller domain. For M different initial conditions this generates M individual samples which match with each other at the boundaries if translated by the domain size in x- or y-direction. This leads to a very flexible arrangement of the samples but has the drawback of frequent recurrence at the boundaries.

The second approach also begins with a computation on the larger domain with boundary Γ^0 but subsequent computations are performed on intermediate domains that extend to the bounds of Γ^0 in directions where a fresh structure is desired at the boundary and are restricted to the bounds of the smaller domain Γ^1 in directions where the structure is to be continued from an already existing neighbor tile by using its stored values and fluxes ϕ_1 and $\mathbf{n} \cdot \nabla \phi_1$ at Γ^1. This requires small modifications of the finite-element formulation in Equations (7) and (8) using only parts of Γ^1 instead of the whole inner boundary. When enough samples are computed to define all N boundary sides the remaining samples are computed on Γ^1 with all sides fixed by boundary conditions from earlier computations.

The most simple setup meeting our design-goal of non-obvious recurrence of boundaries requires five different tiles A_0, B_0, C_0, D_0 and E_0 which can be assembled into a row that matches the same row displaced by a few tiles above and below. Then, five rows of five tiles each form a square which can be continued in all directions indefinitely (see Figure 10c). This setup allows for ten unique boundary sides s_1 through s_{10} (see Figure 11). Inside our 5×5 square of tiles recurrences do not occur in a row, a column, or diagonally, which would not be possible with a smaller number of unique tiles. With only two or three different tiles, repetitions would have to occur at least diagonally (see Figure 10a) and with four different tiles it is only possible to build a unique 4×2 rectangle without diagonal repetitions (see Figure 10b).

On the macroscopic scale our pattern still has obvious repetitions since we need to continue the same square of 5×5 tiles in all directions to form the infinite tiling. To remedy this problem, we generate variants of our five initial tiles A_1, B_1, C_1, D_1 and E_1 which exactly copy the boundaries of their respective prototype with index zero but differ on the inside. Now, in our infinite tiling each tile of a specific prototype is replaced randomly with either the original 0-variant or the new 1-variant of that type. With only two variants per type we already allow for more than 33 million (2^{25}) distinct squares of 5×5 tiles. For our interactive visualization software, we settled with these $M = 10$ tiles due to memory-constraints. However, for printed realizations such as wallpapers, further variants could be computed, making it even harder to spot recurrences.

The initial five tiles must be computed consecutively since their boundaries depend on each other. Starting with A_0 we have no constraints yet and thus the simulation yields four fresh boundary sides. The simulation domain Γ^0 exceeds the region of interest Γ^1 in all four connecting directions because we need to avoid the level lines always being perpendicular to the boundary of Γ^1 (Figure 11a).

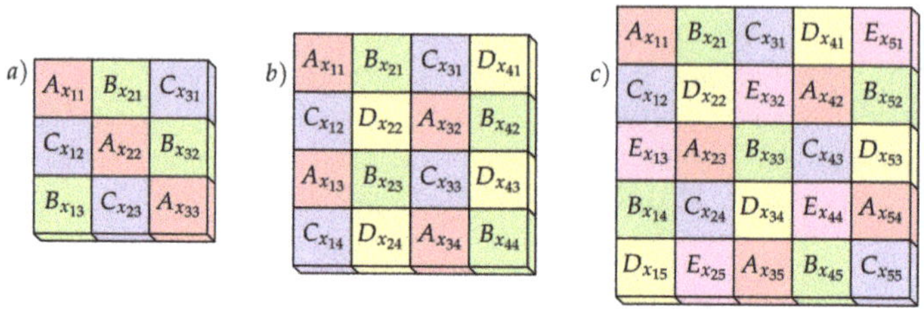

Figure 10. (**a**) Layout of 3 × 3 tiles: diagonal recurrence cannot be avoided. (**b**) Layout of 4 × 4 tiles: the bottom two rows are duplicates of the top two; all other arrangements would yield diagonal recurrence. (**c**) Layout of 5 × 5 tiles: recurrences in non-obvious pattern.

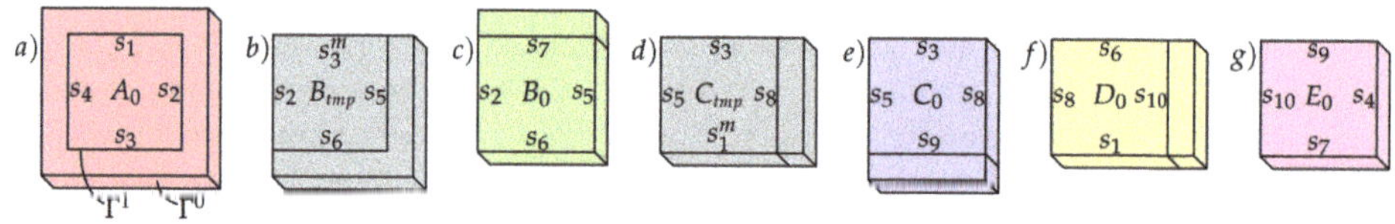

Figure 11. Initial five tiles and their connecting boundary sides. a) Tile A_0 is simulated on Γ^0; b,d) B_{tmp} and C_{tmp} are intermediate steps in preparation of B_0 and C_0, respectively; c,e,f,g) tiles B_0 through E_0 are simulated on domains that extend Γ^1 towards Γ^0 only in directions where fresh boundary data is required, the other boundaries are fixed.

As a next step we simulate a temporary tile B_{tmp} to prepare the run for B_0. Obviously, tile B_0 has to match A_0's right-hand side boundary (s_2 in Figure 11b) at its own left-hand side. However, this is not the only constraint. Our 5 × 5 layout requires that B_0 meets A_0 also in its top-right corner (see Figure 10c). Thus, in our temporary step besides fixing the left boundary we also fix the top boundary with a mirrored version s_3^m of the data from A_0's bottom side. The mirroring is required because the top-right corner of B_0 must conform to the bottom-left corner of A_0. The simulation domain for B_{tmp} exceeds the region of interest only in two directions: bottom and left, where we obtain data for two fresh boundary sides (Figure 11b). Now we are ready to compute our second tile B_0. We fix the right, bottom and left boundaries with data from our previous two simulations and this time leave the top-side unconstrained to obtain a fresh boundary that is only fixed at the two corners where it will match tile A_0 (Figure 11c).

For tile C_0 we require another preliminary run C_{tmp} to fix the bottom-left corner where it meets A_0. We use the mirrored top-side s_1^m of A_0 at C_{tmp}'s bottom to account for that and fix the left and top sides where C_0 shares sides with B_0 and A_0. Only at the left-hand side we obtain a fresh set of boundary-data (Figure 11d). For C_0 we release the bottom-constraint of C_{tmp} and obtain another fresh boundary (Figure 11e). The last tile with a fresh boundary is D_0. All sides, except the right-hand side one, are already constrained by previous computations (Figure 11f). For E_0 all four sides are fully predetermined (Figure 11g). We need to simulate it only for its interior.

All variant-tiles A_1 through E_1 (and further ones if desired) can now be computed in parallel since their boundaries are already known. They are all restricted to Γ^1 and do not require extraction of boundary-values and -fluxes, hence are computationally less expensive than the initial runs.

Supplementary Materials: The following are available online at http://www.mdpi.com/2073-8994/11/4/444/s1, Figure S1: tile1, Figure S2: tile2, Figure S3: tile3, Figure S4: tile4, Video S5: Coarsening of 3D structure, visualized by showing $\phi = 1$ in Ω, Video S6: Coarsening of 3D structure, visualized by showing $\phi = 0.5$ at $z = 0.01, 0.13, 0.25, 0.37, 0.49$, Video S7: Coarsening of 3D structure, visualized by showing $\phi = 0.5$ at $z = 0.01, 0.13, 0.25, 0.37, 0.49$ projected to $z = 0$, Video S8: Journey through space and time in the visualization software in black and white, Video S9: Journey through space and time in the visualization software with colored tiles, Video S10: Rubik's cube.

Author Contributions: F.S. and A.V.; methodology, F.S.; software, F.S. and A.V.; writing—original draft preparation, F.S.; visualization, A.V.; supervision, A.V.; project administration, A.V.; funding acquisition.

Funding: This research was funded by DFG grant number SFB/TRR96 A7. We acknowledge computing resources provided by the Jülich Supercomputing Center under grant number HDR06.

Acknowledgments: We acknowledge preliminary contributions by Folke Post and Lars Ludwig and fruitful discussions with Rainer Backofen and Simon Praetorius.

Conflicts of Interest: The authors declare no conflict of interest. The funders had no role in the design of the study; in the collection, analyses, or interpretation of data; in the writing of the manuscript, or in the decision to publish the results.

References

1. Kohn, R.; Otto, F. Upper bounds on coarsening rates. *Commun. Math. Phys.* **2002**, *229*, 375–395. [CrossRef]
2. Garcke, H.; Niethammer, B.; Rumpf, M.; Weikard, U. Transient coarsening behaviour in the Cahn-Hilliard model. *Acta Mater.* **2003**, *51*, 2823–2830. [CrossRef]
3. Stenger, F.; Voigt, A. Interactive evolution of a bicontinuous structure. *Leonardo* **2019**. [CrossRef]
4. Cahn, J.; Hilliard, J. Free energy of a nonuniform system .1. interfacial free energy. *J. Chem. Phys.* **1958**, *28*, 258–267. [CrossRef]
5. Mullins, W.; Sekerka, R. Morphological stability of a particle growing by diffusion or heat flow. *J. Appl. Phys.* **1963**, *34*, 323–329. [CrossRef]
6. Pego, R.L. Front migration in the nonlinear Cahn-Hilliard equation. *Proc. R. Soc. A* **1989**, *422*, 261–278. [CrossRef]
7. Eyre, D.J. Unconditionally Gradient Stable Time Marching the Cahn-Hilliard Equation. *MRS Online Proc. Libr. Arch.* **1998**, 1686–1712. [CrossRef]
8. Bertozzi, A.L.; Esedoglu, S.; Gillette, A. Inpainting of binary images using the Cahn-Hilliard equation. *IEEE Trans. Image Process.* **2007**, *16*, 285–291. [CrossRef] [PubMed]
9. Glasner, K.; Orizaga, S. Improving the accuracy of convexity splitting methods for gradient flow equations. *J. Comput. Phys.* **2016**, *315*, 52–64. [CrossRef]
10. Diegel, A.; Wang, C.; Wise, S. Stability and convergence of a second-order mixed finite element method for the Cahn–Hilliard equation. *IMA J. Numer. Anal.* **2016**, *36*, 1867–1897. [CrossRef]
11. Backofen, R.; Wise, S.; Salvalaglio, M.; Voigt, A. Convexity splitting in a phase field model for surface diffusion. *Int. J. Numer. Anal. Mod.* **2019**, *16*, 192–209.
12. Boyanova, P.; Do-Quang, M.; Neytcheva, M. Efficient preconditioners for large scale binary Cahn-Hilliard models. *Comput. Meth. Appl. Math.* **2012**, *12*, 1–22. [CrossRef]
13. Praetorius, S.; Voigt, A. Development and analysis of a block-preconditioner for the phase-field crystal equation. *SIAM J. Sci. Comput.* **2015**, *37*, B425–B451. [CrossRef]
14. Vey, S.; Voigt, A. AMDiS: Adaptive multidimensional simulations. *Comput. Vis. Sci.* **2007**, *10*, 57–67. [CrossRef]
15. Witkowski, T.; Ling, S.; Praetorius, S.; Voigt, A. Software concepts and numerical algorithms for a scalable adaptive parallel finite element method. *Adv. Comput. Math.* **2015**, *41*, 1145–1177. [CrossRef]

symmetry

MDPI

Article

New Numerical Method for the Rotation form of the Oseen Problem with Corner Singularity

Viktor A. Rukavishnikov [1,*] and Alexey V. Rukavishnikov [2]

1 Computing Center of Far-Eastern Branch, Russian Academy of Sciences, Kim-Yu-Chen Str. 65, Khabarovsk 680000, Russia

2 Institute of Applied Mathematics of Far-Eastern Branch, Khabarovsk Division, Russian Academy of Sciences, Dzerzhinsky Str. 54, Khabarovsk 680000, Russia; 78321a@mail.ru

* Correspondence: vark0102@mail.ru; Tel.: +8-421-257-2620

Received: 18 November 2018; Accepted: 12 December 2018; Published: 5 January 2019

Abstract: In the paper, a new numerical approach for the rotation form of the Oseen system in a polygon Ω with an internal corner ω greater than 180° on its boundary is presented. The results of computational simulations have shown that the convergence rate of the approximate solution (velocity field) by weighted FEM to the exact solution does not depend on the value of the internal corner ω and equals $\mathcal{O}(h)$ in the norm of a space $W^1_{2,\nu}(\Omega)$.

Keywords: Oseen problem; corner singularity; weighted finite element method; preconditioning

1. Introduction

Many mathematical models of natural processes are described by the boundary value problems for systems of partial differential equations with a singularity. The singularity of the solution to such systems in the two-dimensional closed domain Ω may be due to the degeneration of initial data, to the presence of reentrant corners on a boundary, or to internal features of the solution. The boundary value problem has a strong singularity if its solution does not belong to the Sobolev space $W^1_2(\Omega)$. In short, the Dirichlet integral from the solution diverges. In the case when the solution belongs to the space $W^1_2(\Omega)$, but it does not belong to the $W^2_2(\Omega)$, a boundary value problem is called weakly singular. The generalized solution of a boundary value problem in the two-dimension domain with a boundary containing an initial angle ω belongs to the space $W^{1+\alpha-\epsilon}_2(\Omega)$, where $0.25 \leq \alpha < 1$ for $\pi < \omega \leq 2\pi$ and ϵ is an arbitrary positive real number. Therefore, the approximate solution produced by the classical finite difference or finite element methods converges to an exact one no faster than at the $\mathcal{O}(h^\alpha)$ rate [1].

For the boundary value problem with singularity, there are various numerical approaches founded on the separation of singular and regular components of the generalized solution, on mesh refinement toward singularity points, and on the multiplicative identification of singularities. These methods slow down the convergence rate of the approximate solution to an exact one or to the significant complication of the finite element scheme, which in total influences the computational process speed and accuracy of the result.

In reference [2], we suggested to define the solution of the boundary value problem with weak or strong singularity as an R_ν-generalized one in the weighted Sobolev space or set. Relying on this approach, numerical methods were created with a convergence rate independent of the value (size) of a singularity. In the papers [3–5] for the boundary value problems with a strong singularity, the weighted finite element method (FEM) and the weighted edge-based FEM were built. The approximate solution converges to an exact one with the second and first order rates (under the mesh step h) in the norms of the weighted Lebesgue and Sobolev spaces, respectively. In references [6,7], a weighted FEM for the

Lame system in a domain with the reentrant corner on the boundary was built. The rate of convergence is equal to $\mathcal{O}(h)$ and independent of the size of a reentrant corner.

We study the incompressible Navier–Stokes equations in the two-dimensional polygonal domain Ω with one internal corner greater than 180° on its boundary. The nonlinearity in this system can be written in several equivalent forms. For one case, if we regard these equations in the velocity field and kinematic pressure variables, then this leads to the convection form of nonlinear terms. For another case, if we consider these equations in the velocity field and total pressure variables, then it gives nonlinear terms in the rotation form. In order to meet the non-stationary incompressible system, we must be able to find the solution of a steady linearized one. The stationary Navier–Stokes system we can linearize in different manners. We use a scheme that is based on Picard's iterative procedure (see [8] and the references therein). Starting with an arbitrary vector as a velocity field, which satisfies the law of conservation of mass, Picard's iterative procedure forms the sequence of solutions of the corresponding linear Oseen system. We note that linearizations of convection and rotation forms of nonlinear terms tend to the systems of linear algebraic equations with various features. In the paper, we study the Oseen system in the rotation form. The fact is that the rotation form allows us (using a skew-symmetric of the resulting matrix) to construct a Schur complement preconditioner, which is acceptable to all parameters of the Oseen problem and becomes more effective for large Reynolds numbers (see [9] and the references therein). For the convection form of the Oseen problem, this is not so.

As usual, to solve a fluid problem, the explorer has freedom and can construct a method in different manners by selecting various discretization algorithms for the system of linear algebraic equations. There are many opportunities to solve the considered system. The researcher can select various finite difference, finite volume, or finite element methods. However, the chosen method is effective if it gives the best result in terms of the convergence rate under certain restrictions on the input data and geometric singularities of the domain Ω.

In the paper, we consider a special case, where Ω is a polygon with one internal corner greater than 180° on its boundary. The flow of the viscous fluid in a δ-neighborhood of a reentrant angle was studied in [10]. It is not a secret that the velocity field and pressure, as a weak solution of a problem for the domain with corner singularity, do not belong to Sobolev spaces $\mathbf{W}_2^2(\Omega)$ and $W_2^1(\Omega)$, respectively [11]. Therefore, the rate of convergence of the approximate solution to an exact one is equal to $\mathcal{O}(h^\alpha)$, $\alpha < 1$, in the norm of standard and weighted Sobolev spaces (see [12] and the references therein) for different classical finite difference and finite element methods. Earlier, for the Stokes problem, we defined the R_ν-generalized solution; in [13], we formulated and proved the weighted LBBinequality (inf-sup condition [14]); and in [15], we showed the advantage of our method over classical approaches.

The aim of the paper is to present a new numerical approach for the rotation form of the Oseen problem using (see [16]) a mass conservation space pair; to show that the rate of convergence of the approximate solution to an exact one (the velocity field) is equal to $\mathcal{O}(h)$ for all considered sizes of the internal corner greater than 180° on the boundary in the norm of the space $\mathbf{W}^1_{2,\nu}(\Omega_k)$; so that this rate is much better than if using the classical finite difference or finite element methods.

The article consists of six sections. Section 2 is devoted to the definition of the R_ν-generalized solution for the rotation form of the Oseen system in a domain Ω with one internal corner greater than 180° on its boundary. In Section 3, we construct the presented FEM. The iterative algorithm for the resulting system of linear algebraic equations is built in Section 4. In Section 5, we discuss the numerical results of computational experiments. Necessary conclusions are made in Section 6.

2. R_ν-Generalized Solution of the Oseen Problem

Let $\mathbf{x} = (x_1, x_2)$ be an element of the Euclidean space $\mathbf{R}^2$, where $\|\mathbf{x}\| = \left(x_1^2 + x_2^2\right)^{1/2}$ and $d\mathbf{x} = dx_1\, dx_2$ are the norm and measure of $\mathbf{x}$, respectively. Denote by Ω a bounded domain in $\mathbf{R}^2$. Let Γ and $\bar{\Omega}$ be the boundary and closure of Ω, respectively, where $\bar{\Omega} = \Omega \cup \Gamma$.

At first, we write incompressible Navier–Stokes equations in such a form: find a velocity field $\mathbf{u}(\mathbf{x},t) = (u_1(\mathbf{x},t), u_2(\mathbf{x},t))$ and a kinematic pressure $p(\mathbf{x},t)$ from:

$$\frac{\partial \mathbf{u}}{\partial t} - \bar{\nu}\triangle\mathbf{u} + (\mathbf{u}\cdot\nabla)\mathbf{u} + \nabla p = \mathbf{f} \quad \text{and} \qquad \text{div}\,\mathbf{u} = \mathbf{0} \quad \text{in} \qquad \Omega\times(0,T], \tag{1}$$

with given force field $\mathbf{f} = (f_1, f_2)$ and viscosity $\bar{\nu} = \frac{1}{Re} > 0$. Let $\triangle$, div , and ∇ be the Laplace, divergence, and gradient operators in $\mathbf{R}^2$, respectively. The equations in (1) are the convection form of Navier–Stokes equations.

We supplement the system (1) with a boundary and initial conditions:

$$\mathbf{u} = \mathbf{g} \quad \text{on} \quad \Gamma\times(0,T], \qquad \mathbf{u}(\mathbf{x},0) = \mathbf{u}^0(\mathbf{x}) \quad \text{in} \quad \Omega, \tag{2}$$

where $\mathbf{g} = (g_1, g_2)$ is given vector function on Γ and $\mathbf{u}^0(\mathbf{x}) = (u_1^0(\mathbf{x}),\, u_2^0(\mathbf{x}))$ — in Ω.

We introduce the following notation:

$$\mathbf{v}\cdot\mathbf{u} = \sum_{i=1}^{2} u_i v_i, \qquad \text{curl}\,\mathbf{u} = -\frac{\partial u_1}{\partial x_2} + \frac{\partial u_2}{\partial x_1}, \qquad a\times\mathbf{u} = \begin{pmatrix} -au_2 \\ au_1 \end{pmatrix}.$$

We have a formal equality:

$$\nabla(\mathbf{u}\cdot\mathbf{v}) + (\,\text{curl}\,\mathbf{u})\times\mathbf{v} + (\,\text{curl}\,\mathbf{v})\times\mathbf{u} = (\mathbf{u}\cdot\nabla)\mathbf{v} + (\mathbf{v}\cdot\nabla)\mathbf{u}. \tag{3}$$

If $\mathbf{u} = \mathbf{v}$ in (3), then we have a relation:

$$(\,\text{curl}\,\mathbf{v})\times\mathbf{v} + \frac{1}{2}\nabla\mathbf{v}^2 = (\mathbf{v}\cdot\nabla)\mathbf{v}. \tag{4}$$

Let $P = p + \dfrac{1}{2}\mathbf{u}^2$, using (4), for vector function $\mathbf{u}$; we get the rotation form of the Navier–Stokes system for an incompressible flow:

$$\frac{\partial \mathbf{u}}{\partial t} - \bar{\nu}\triangle\mathbf{u} + (\,\text{curl}\,\mathbf{u})\times\mathbf{u} + \nabla P = \mathbf{f} \ \text{and} \quad \text{div}\,\mathbf{u} = \mathbf{0} \ \text{in} \ \Omega\times(0,T]. \tag{5}$$

We supplement the system (5) with the boundary and initial conditions (2). Using implicit time integration of (5) compared to explicit methods reduces accuracy, stability, and flexibility in selecting the step size for a time variable.

In our research, on each time level, we solve the following system of equations:

$$-\bar{\nu}\triangle\mathbf{u} + \ \text{curl}\,\mathbf{u}\times\mathbf{u} + \alpha\,\mathbf{u} + \nabla P = \mathbf{f} \ \text{and} \quad \text{div}\,\mathbf{u} = \mathbf{0} \quad \text{in} \quad \Omega, \tag{6}$$

$$\mathbf{u} = \mathbf{g} \qquad \text{on} \qquad \Gamma, \tag{7}$$

and parameter α is a known positive constant.

The system (6) and (7) is nonlinear due to the fact that there is a rotation term curl $\mathbf{u}\times\mathbf{u}$ in the first Equation (6). This term and the system as a whole we linearized by Picard's iterative procedure (see [8] and the references therein).

At each iteration, we need to solve the following problem:

$$-\bar{\nu}\triangle\mathbf{u} + w\times\mathbf{u} + \alpha\,\mathbf{u} + \nabla P = \mathbf{f}, \ \text{and} \quad \text{div}\,\mathbf{u} = \mathbf{0} \quad \text{in} \quad \Omega, \tag{8}$$

$$\mathbf{u} = \mathbf{g} \qquad \text{on} \qquad \Gamma, \tag{9}$$

which is called the Oseen system in a rotation form, where $w =$ curl $\mathbf{U}$ and $\mathbf{U}$ is some approximation to $\mathbf{u}$.

The linearization of convection and rotation forms of nonlinear terms tends to the systems of linear algebraic equations with various features. In the paper, we study the Oseen system in the rotation form. The fact is that the rotation form allows us (using a skew-symmetric of the resulting matrix) to construct the Schur complement preconditioner, which is acceptable to all parameters of the Oseen problem and becomes more effective when $\bar{\nu} \to 0$ (see [9] and the references therein). For the convection form of the Oseen problem, this is not so.

We note that for the linearized system (8) and (9), the laws of the conservation of momentum and mass remain valid.

In the paper, we consider a special case, where Ω is a bounded non-convex polygonal domain with one internal corner greater than 180° on Γ. Let its vertex be located at the origin. We define an R_ν-generalized solution of the Oseen problem (8) and (9) with a corner singularity and construct the weighted FEM. We demonstrate the advantage of the proposed approach over the classical finite element methods for all sizes of the reentrant corner.

Let $\Omega'_\delta = \{\mathbf{x} \in \bar{\Omega} : \|\mathbf{x}\| \le \delta, \delta \in (0,1)\}$ be a part of a δ-neighborhood, with a vertex located at the origin, which is in $\bar{\Omega}$. Denote by $\rho(\mathbf{x})$ a weight function: $\rho(\mathbf{x}) = \begin{cases} \|\mathbf{x}\|, \mathbf{x} \in \Omega'_\delta, \\ \delta \ \ , \mathbf{x} \in \bar{\Omega} \setminus \Omega'_\delta. \end{cases}$

Let $D^m v(\mathbf{x}) = \frac{\partial^{|m|} v(\mathbf{x})}{\partial x_1^{m_1} \partial x_2^{m_2}}$ be the mth order generalized derivatives of a function $v(\mathbf{x})$ in Ω, where $|m| = m_1 + m_2, m_i$, nonnegative integers. For the function $v(\mathbf{x})$, we define the following inequalities:

$$\int\limits_{\Omega \setminus \Omega'_\delta} \rho^{2\alpha} v^2 d\mathbf{x} \ge C_1 > 0, \tag{10}$$

$$|D^m v(\mathbf{x})| \le C_2 \Big(\frac{\delta}{\rho(\mathbf{x})}\Big)^{\alpha+m} \quad \text{for } \mathbf{x} \in \Omega'_\delta \text{ and } m = 0, 1, \tag{11}$$

where $\alpha > 0$ and constant $C_2 > 0$ do not depend on m and α.

Denote by $L_{2,\alpha}(\Omega)$ a space of functions $v(\mathbf{x})$, such that:

$$\|v\|_{L_{2,\alpha}(\Omega)} = \Big(\int\limits_\Omega \rho^{2\alpha} v^2 d\mathbf{x}\Big)^{1/2} < \infty.$$

If $\mathbf{w} = (w_1, w_2)$ is a vector function, then we define the weighted vector function space $\mathbf{L}_{2,\alpha}(\Omega)$ with a norm $\|\mathbf{w}\|_{\mathbf{L}_{2,\alpha}(\Omega)} = \Big(\|w_1\|^2_{L_{2,\alpha}(\Omega)} + \|w_2\|^2_{L_{2,\alpha}(\Omega)}\Big)^{1/2}$.

Further, denote by $L_{2,\alpha}(\Omega, \delta), \alpha > 0$, a set of elements $v(\mathbf{x})$ from the $L_{2,\alpha}(\Omega)$ space for which Inequalities (10) and (11) (the case $m = 0$) are valid with a bounded $L_{2,\alpha}(\Omega)$ norm. Let $L^0_{2,\alpha}(\Omega, \delta)$ be a subset of functions $v(\mathbf{x})$, such that $L^0_{2,\alpha}(\Omega, \delta) = \{v \in L_{2,\alpha}(\Omega, \delta) : \int\limits_\Omega \rho^\alpha v d\mathbf{x} = 0\}$. If $\mathbf{w} = (w_1, w_2)$ is a vector function, then we define a set $\mathbf{L}_{2,\alpha}(\Omega, \delta) = \{\mathbf{w} : w_i \in L_{2,\alpha}(\Omega, \delta)\}$ with a bounded $\mathbf{L}_{2,\alpha}(\Omega)$ norm.

Let $W^1_{2,\alpha}(\Omega)$ be a weighted space of functions $v(\mathbf{x})$, such that:

$$\|v\|_{W^1_{2,\alpha}(\Omega)} = \Big(\sum_{|m| \le 1} \|\rho^\alpha |D^m v|\|^2_{L_2(\Omega)}\Big)^{1/2} < \infty.$$

If $\mathbf{w} = (w_1, w_2)$ is a vector function, then we denote by $\mathbf{W}^1_{2,\alpha}(\Omega)$ the weighted vector function space with a norm $\|\mathbf{w}\|_{\mathbf{W}^1_{2,\alpha}(\Omega)} = \Big(\|w_1\|^2_{W^1_{2,\alpha}(\Omega)} + \|w_2\|^2_{W^1_{2,\alpha}(\Omega)}\Big)^{1/2}$.

Let $W^1_{2,\alpha}(\Omega,\delta), \alpha > 0$, be a set of functions $v(\mathbf{x})$ from the space $W^1_{2,\alpha}(\Omega)$, that meet the conditions (10) and (11) with a bounded $W^1_{2,\alpha}(\Omega)$ norm. We denote by $\overset{o}{W}{}^1_{2,\alpha}(\Omega,\delta)(\overset{o}{W}{}^1_{2,\alpha}(\Omega,\delta) \subset W^1_{2,\alpha}(\Omega,\delta))$ a closure, with respect to the $W^1_{2,\alpha}(\Omega)$ norm, of the set of infinitely-differentiable functions with compact support in Ω that meet the inequalities (10) and (11). Then, we denote by $W^{1/2}_{2,\alpha}(\Gamma,\delta)$ the set of functions $\theta(\mathbf{x})$ on $\Gamma : \theta(\mathbf{x}) \in W^{1/2}_{2,\alpha}(\Gamma,\delta)$, if there exists a function $\Theta(\mathbf{x})$ from the set $W^1_{2,\alpha}(\Omega,\delta)$, such that $\Theta(\mathbf{x})|_\Gamma = \theta(\mathbf{x})$ and $\|\theta\|_{W^{1/2}_{2,\alpha}(\Gamma,\delta)} = \inf\limits_{\Theta|_\Gamma=\theta} \|\Theta\|_{W^1_{2,\alpha}(\Omega,\delta)}$.

If $\mathbf{w} = (w_1, w_2)$ is a vector function, then we define a set $\mathbf{W}^1_{2,\alpha}(\Omega,\delta) = \{\mathbf{w} : w_i \in W^1_{2,\alpha}(\Omega,\delta)\}$ with a norm of space $\mathbf{W}^1_{2,\alpha}(\Omega)$. Similarly, we define the set $\overset{o}{\mathbf{W}}{}^1_{2,\alpha}(\Omega,\delta)$ of vector functions in Ω and $\mathbf{W}^{1/2}_{2,\alpha}(\Gamma,\delta)$, on Γ.

Let known functions w, $\mathbf{f} = (f_1, f_2)$ and $\mathbf{g} = (g_1, g_2)$ in (8) and (9) meet the following conditions:

$$w \in L_{2,\gamma}(\Omega,\delta), \quad \mathbf{f} \in \mathbf{L}_{2,\gamma}(\Omega,\delta), \quad \mathbf{g} \in \mathbf{W}^{1/2}_{2,\gamma}(\Gamma,\delta), \ \gamma \geq 0. \tag{12}$$

Bilinear and linear forms are as follows:

$$a(\mathbf{u}_\nu, \mathbf{v}) = \int\limits_\Omega \Big[\bar{\nu}\nabla\mathbf{u}_\nu \cdot \nabla(\rho^{2\nu}\mathbf{v}) + \rho^{2\nu}(w \times \mathbf{u}_\nu) \cdot \mathbf{v} + \alpha\rho^{2\nu}\mathbf{u}_\nu \cdot \mathbf{v}\Big] d\mathbf{x},$$

$$b(\mathbf{v}, P_\nu) = -\int\limits_\Omega P_\nu \ \mathrm{div}\,(\rho^{2\nu}\mathbf{v}) d\mathbf{x}, \ c(\mathbf{u}_\nu, q) = -\int\limits_\Omega \rho^{2\nu} q \ \mathrm{div}\,\mathbf{u}_\nu \, d\mathbf{x}, \ l(\mathbf{v}) = \int\limits_\Omega \rho^{2\nu}\,\mathbf{f} \cdot \mathbf{v} d\mathbf{x}.$$

Definition 1. *The pair* $(\mathbf{u}_\nu(\mathbf{x}), P_\nu(\mathbf{x})) \in \mathbf{W}^1_{2,\nu}(\Omega,\delta) \times L^0_{2,\nu}(\Omega,\delta)$ *is called the* R_ν*-generalized solution for an Oseen system in the rotation form (8) and (9) such that for all pairs* $(\mathbf{v}(\mathbf{x}), q(\mathbf{x})) \in \overset{o}{\mathbf{W}}{}^1_{2,\nu}(\Omega,\delta) \times L^0_{2,\nu}(\Omega,\delta)$, *the equalities:*

$$\begin{aligned} a(\mathbf{u}_\nu, \mathbf{v}) + b(\mathbf{v}, P_\nu) &= l(\mathbf{v}), \\ c(\mathbf{u}_\nu, q) &= 0 \end{aligned}$$

hold, where functions w, $\mathbf{f}$ *and* $\mathbf{g}$ *satisfy the conditions (12) and* $\nu \geq \gamma$.

Note that the bilinear and linear forms in the definition of an R_ν-generalized solution include a weight function $\rho(\mathbf{x})$. The introduction of the weight function into integral identities suppresses the influence of the singularity in the solution and ensures that $\mathbf{u}_\nu$ and P_ν belong to the weighted sets $\mathbf{W}^2_{2,\nu}(\Omega,\delta)$ and $W^1_{2,\nu}(\Omega,\delta)$, respectively. This property of the R_ν-generalized solution allows one to construct a finite element scheme with a $\mathcal{O}(h)$ rate. This rate is significantly higher than in the classical finite element method for the Oseen problem in a polygonal domain with the internal corner greater than 180° on the boundary.

3. The Weighted Finite Element Scheme

Now, we construct a finite element scheme for an Oseen problem in the rotation form (8) and (9) based on the definition of an R_ν-generalized solution.

We would like to use the finite element space pair, which satisfies the law of mass conservation not in the weak (like the well-known Taylor–Hood (TH) element pair [14]), but in the strong sense. The fact is that the implementation of the mass conservation law in a weak sense combines pressure and velocity field errors and does not eliminate possible instabilities [17]. In the paper, we apply the Scott–Vogelius (SV) element pair [16] that will help us to obtain strong mass conservation of the approximate solution.

First, we divide $\overline{\Omega}$ into a finite quantity of triangles L_i, which we call macro-elements. The set of elements L_i represents a quasi-uniform (see [1]) triangulation T_h of $\overline{\Omega}$. Then, we divide each macro-element $L_i \in T_h$ into three triangles K_{i_j} using the barycenter of L_i. Thus, we construct a triangulation Υ_h, which is based on a barycenter refinement of a triangulation T_h. Denote by Ω_h the set of resulting triangles (which are called finite elements) with sides of order h, i.e., $\Omega_h = \bigcup\limits_{K_{i_j} \in \Upsilon_h} K_{i_j} = \bigcup\limits_{L_i \in T_h} \left(\bigcup\limits_{j=1}^{3} K_{i_j} \right) = \bigcup\limits_{L_i \in T_h} L_i$.

Let A_m and B_l be vertices and midpoints of the finite element sides $K_{i_j} \in \Upsilon_h$, respectively. Then, for the components of a velocity field and pressure, we define sets of nodes G and H, respectively, such that $G = G_\Omega \bigcup G_\Gamma = \{A_m \cup B_l\}$, where G_Ω is a totality of Υ_h nodes in Ω, G_Γ, on Γ, and $H = \{C_k\}$, where C_k coincide with a node A_m on the appropriate element $K_{i_j} \in \Upsilon_h$ (see Figure 1).

Now, we define spaces of the SV element pair. The space X_h, for the components of the velocity field, coincides with the corresponding space of degree two of the THelement pair, i.e., $X^h = \{w^h \in C(\Omega) : w^h|_{K_{i_j}} \in P_2(K_{i_j}), \forall K_{i_j} \in \Upsilon_h\}$ and for a velocity field $\mathbf{X}^h = X^h \times X^h$. The space Y^h, for the pressure, differs from the corresponding space degree one of the TH element pair by the fact that it is discontinuous in Ω, i.e., $Y^h = \{y^h \in L_2(\Omega) : y^h|_{K_{i_j}} \in P_1(K_{i_j}), \forall K_{i_j} \in \Upsilon_h, \int\limits_\Omega y^h d\mathbf{x} = 0\}$.

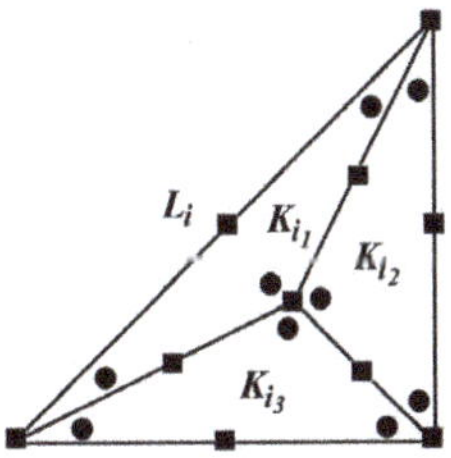

Figure 1. The macro-element L_i: squares and dots are the velocity and pressure nodes on K_{i_j}, $j = 1, 2, 3$, respectively.

The SV element pair has an important property, namely div $\mathbf{X}^h \subset Y^h$. This means that there exists a function $y^h \in Y^h$ equal to div $\mathbf{w}^h$ such that: from the condition for performing mass conservation in a weak sense, i.e., $\int\limits_\Omega \text{div}\, \mathbf{w}^h \psi^h d\mathbf{x} = 0 \; \forall \psi^h \in Y^h$, we get a pointwise mass conservation, i.e., $\| \text{div}\, \mathbf{w}^h \|_{L_2(\Omega)} = 0$. Moreover, in [18], it was established that spaces of the SV element pair before us satisfy the Ladyzhenskaya–Babuška–Brezzi condition. Note, that approximations obtained using the TH element, pair unlike the SV element pair, in general, do not achieve pointwise mass conservation.

Then, we define the weighted basis functions and describe a special finite element method for the Oseen system in the rotation form (8) and (9). For components of the velocity field, for each node $M_k \in G_\Omega$, we will match a function:

$$\Phi_k(\mathbf{x}) = \rho^{\nu^\circ}(\mathbf{x}) \cdot \varphi_k(\mathbf{x}), \; k = 0, 1, \ldots,$$

where $\varphi_k(\mathbf{x}) \in X^h, \varphi_k(M_j) = \begin{cases} 1, k = j, \\ 0, k \neq j, \end{cases}$ $k, j = 0, 1, \ldots;$ ν° is a parameter.

We define a set V^h, for components of the velocity field, such that for any velocity field $\mathbf{v}^h = (v_1^h, v_2^h), v_i^h \in V^h$, we have:

$$v_1^h(\mathbf{x}) = \sum_k d_{2k}\, \Phi_k(\mathbf{x}), \; v_2^h(\mathbf{x}) = \sum_k d_{2k+1}\, \Phi_k(\mathbf{x}), \tag{13}$$

where $d_l = \rho^{-\nu^\circ}(M_{[l/2]})\, \tilde{d}_l$.

Let V_0^h be a subset in V^h such that $V_0^h = \{w^h \in V^h : w^h(M_k)|_{M_k \in G_\Gamma} = 0\}$. Moreover, we define velocity field sets $\mathbf{V}^h = V^h \times V^h$ and $\mathbf{V}_0^h = V_0^h \times V_0^h$.

For the pressure, for each node $N_l \in H$, we will match a function:

$$\Theta_m(\mathbf{x}) = \rho^{\mu^\diamond}(\mathbf{x}) \cdot \theta_m(\mathbf{x}), \ m = 0, 1, \ldots,$$

where $\theta_m(\mathbf{x}) \in Y^h, \theta_m(N_j) = \begin{cases} 1, m = j, \\ 0, m \neq j, \end{cases}$ $m, j = 0, 1, \ldots;$ $\mu^\diamond$ is a parameter.

Then, we define a set Q^h, for the pressure, such that for any $q^h \in Q^h$, we have:

$$q^h(\mathbf{x}) = \sum_m e_m \, \Theta_m(\mathbf{x}), \tag{14}$$

where $e_m = \rho^{-\mu^\diamond}(N_m)\, \tilde{e}_m$.

Remark 1. *The coefficients d_j and e_i in (13) and (14) are defined as a solution of a system (17) (see below).*

Remark 2. *The following embedding of sets is valid:*

$$\mathbf{V}^h \subset \mathbf{W}^1_{2,\nu}(\Omega_h, \delta), \mathbf{V}_0^h \subset \overset{o}{\mathbf{W}^1_{2,\nu}}(\Omega_h, \delta), \ Q^h \subset L^0_{2,\nu}(\Omega_h, \delta).$$

Definition 2. *The pair $(\mathbf{u}_\nu^h(\mathbf{x}), P_\nu^h(\mathbf{x})) \in \mathbf{V}^h \times Q^h$ is called an approximate R_ν-generalized solution for an Oseen system in the rotation form (8) and (9) obtained by the weighted FEM if the equalities:*

$$a(\mathbf{u}_\nu^h, \mathbf{v}^h) + b(\mathbf{v}^h, P_\nu^h) = l(\mathbf{v}^h), \tag{15}$$

$$c(\mathbf{u}_\nu^h, q^h) = 0 \tag{16}$$

hold for any pair $(\mathbf{v}^h(\mathbf{x}), q^h(\mathbf{x})) \in \mathbf{V}_0^h \times Q^h$, where $\mathbf{u}_\nu^h = (u_{\nu,1}^h, u_{\nu,2}^h)$ and $\omega \in L_{2,\gamma}(\Omega, \delta), \mathbf{f} \in \mathbf{L}_{2,\gamma}(\Omega, \delta), \mathbf{g} \in \mathbf{W}^{1/2}_{2,\gamma}(\Gamma, \delta), \nu \geq \gamma$.

Thus, we construct a weighted FEM to find an R_ν-generalized solution for the rotation form of the Oseen problem (8) and (9).

Then, using (15) and (16), we get a system of linear algebraic equations:

$$\mathbf{Ad} + \mathbf{Be} = \omega, \qquad \mathbf{C}^T\mathbf{d} = \mathbf{z}, \tag{17}$$

where $\mathbf{d} = (d_0, d_2, d_4, \ldots, d_1, d_3, d_5, \ldots)^T, \mathbf{e} = (e_0, e_1, e_2, \ldots)^T, \omega = \mathbf{F}^h, \mathbf{z} = \mathbf{0}$.

4. Iterative Algorithm

Now, we present an iterative procedure for solving the system of equations (17). Note that the system (17), which needs to be solved, has a large dimension, and moreover, its matrix is sparse. Finding the solution of the system by the direct method is not possible, so that we will construct a convergent iterative process of the following type [19]:

(1) Let $(\mathbf{d}^0, \mathbf{e}^0)$ be an initial guess for the system (17). We iterate $(n = 0, 1, 2, \ldots)$ until the stopping condition is fulfilled;
(2) Compute $\mathbf{d}^{n+1} = \mathbf{d}^n + \hat{\mathbf{A}}^{-1}(\omega - \mathbf{Ad}^n - \mathbf{Be}^n)$;
(3) Find $\mathbf{e}^{n+1} = \mathbf{e}^n + \hat{\mathbf{S}}^{-1}(\mathbf{C}^T\mathbf{d}^{n+1} - \mathbf{z})$;

where $\hat{\mathbf{A}}$ is a preconditioning matrix to $\mathbf{A}$ and $\hat{\mathbf{S}}$ is a preconditioning matrix to $\mathbf{S} = \mathbf{C}^T\mathbf{A}^{-1}\mathbf{B}$, which is called the Schur complement matrix. Next, we describe the process of constructing preconditioning matrices $\hat{\mathbf{A}}$ and $\hat{\mathbf{S}}$.

At first, we build a preconditioner $\hat{\mathbf{A}}$ applying an incomplete $\mathbf{LU}$ factorization, where $\mathbf{L}$ and $\mathbf{U}$ are low unitriangular and upper triangular matrices respectively. At each iteration in Item 2, we employ the GMRES(l) method (see [20]) as the solution of a problem $\mathbf{Av} = \mathbf{s}$ with the left preconditioner $\hat{\mathbf{A}}$. The method is designed so that it approximates the solution in an lth order Krylov subspace. In our research, the dimension of a Krylov subspace is equal to 10; so that if $\mathbf{r}_0 = \hat{\mathbf{A}}^{-1}(\mathbf{s} - \mathbf{Av})$, then the Arnoldi procedure will build an orthonormal basis of the subspace: $\text{Span}\{\mathbf{r}_0, (\hat{\mathbf{A}}^{-1}\mathbf{A})^1\mathbf{r}_0, \ldots, (\hat{\mathbf{A}}^{-1}\mathbf{A})^9\mathbf{r}_0\}$.

Secondly, we build an intermediate matrix $\tilde{\mathbf{S}}$ to $\hat{\mathbf{S}}$. The matrix $\tilde{\mathbf{S}}$ represents a mass matrix $\mathbf{M}_P^{\overline{\nu},\mu^\diamond,\nu}$ of a special view, such that on all elements $K \in \Upsilon_h$:

$$(\mathbf{M}_P^{\overline{\nu},\mu^\diamond,\nu})_{ij} = \frac{1}{\overline{\nu}} \int\limits_K \rho^{2(\nu+\mu^\diamond)}\, \theta_j(\mathbf{x})\, \theta_i(\mathbf{x}) d\mathbf{x},\ \theta_j(\mathbf{x}), \theta_i(\mathbf{x}) \in Y^h, j, i = 0, 1, \ldots.$$

After that, we determine a matrix $\bar{\mathbf{S}}$, which is equal to a diagonal matrix $\bar{\mathbf{M}}_P^{\overline{\nu},\mu^\diamond,\nu}$ with elements $\left(\bar{\mathbf{M}}_P^{\overline{\nu},\mu^\diamond,\nu}\right)_{ii} = \sum\limits_k \left(\mathbf{M}_P^{\overline{\nu},\mu^\diamond,\nu}\right)_{ik}$. In other words, $\left(\bar{\mathbf{S}}\right)_{ii} = \sum\limits_k \left(\tilde{\mathbf{S}}\right)_{ik}$. It is known (see [9] and the references therein) that such diagonal lumping $\bar{\mathbf{S}}$ is a good preconditioner to the initial matrix $\tilde{\mathbf{S}}$.

Therefore, in order to determine the vector $\Psi^\star := \hat{\mathbf{S}}^{-1}\chi$, at each iteration of Item 3, we must find a solution to the following internal procedure: (1) $\phi^0 = \mathbf{0}$; (2) $\phi^m = \phi^{m-1} + \bar{\mathbf{S}}^{-1}(\chi - \tilde{\mathbf{S}}\phi^{m-1})$ $(m = 1, \ldots, M)$; (3) $\Psi^\star = \phi^M$.

We apply the GMRES(5) method, where $\text{Span}\{\bar{\mathbf{r}}, (\bar{\mathbf{S}}^{-1}\tilde{\mathbf{S}})^1\bar{\mathbf{r}}, \ldots, (\bar{\mathbf{S}}^{-1}\tilde{\mathbf{S}})^4\,\bar{\mathbf{r}}\}$, and $\bar{\mathbf{r}} = \bar{\mathbf{S}}^{-1}(\chi - \tilde{\mathbf{S}}\phi^{m-1})$.

5. Numerical Experiments

Now, we present numerical results for the Oseen system in the rotation form (8) and (9) and show the advantage of the proposed method.

Let $\Omega_k = (-l;\, l) \times (-l;\, l) \setminus \bar{G}_k$ be a polygon with one internal corner greater than 180° on Γ_k whose vertex is at the origin. We will consider the following sizes of the reentrant corner: $\omega_k = \frac{2^k+1}{2^k}\pi$, $k = 1, 2, 3$. The triangulation Υ_h (see Section 3) of each $\bar{\Omega}_k, k = 1, 2, 3$ and $l = 1$ we present in Figure 2.

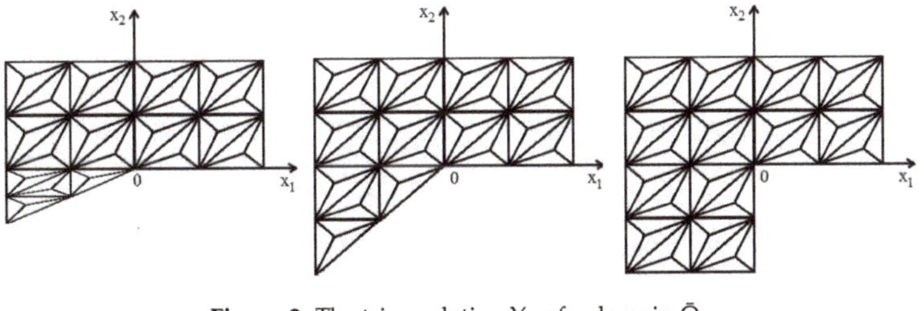

Figure 2. The triangulation Υ_h of a domain $\bar{\Omega}_k$.

In a test problem, we consider the solution of the problem (8), (9), which has a singularity in a neighborhood of a point located at the origin. Let $\alpha = \bar{\nu} = 1, w = b \cdot$ curl $\mathbf{u}, b = 0.95$, and for each corner ω_k in polar coordinates $(r,\ \varphi)$, we have an auxiliary function:

$$\Psi_k(\varphi) = \frac{\sin((1+\lambda_k)\varphi)\cos(\lambda_k\omega_k)}{1+\lambda_k} - \frac{\sin((1-\lambda_k)\varphi)\cos(\lambda_k\omega_k)}{1-\lambda_k} + \cos((1-\lambda_k)\varphi) - \cos((1+\lambda_k)\varphi).$$

Then, the exact solution $\mathbf{u} = (u_1, u_2)$ and P of the problem (8) and (9) for each corner $\omega_k, k = 1, 2, 3$, in polar coordinates has the following form:

$$u_1(r,\varphi) = r^{\lambda_k} \cdot ((\lambda_k+1)\sin\varphi \cdot \Psi_k(\varphi) + \cos\varphi \cdot \Psi_k'(\varphi)),$$
$$u_2(r,\varphi) = r^{\lambda_k} \cdot (\sin\varphi \cdot \Psi_k'(\varphi) - (\lambda_k+1)\cos\varphi \cdot \Psi_k(\varphi)),$$
$$P(r,\varphi) = r^{\lambda_k - 1} \cdot \frac{(\lambda_k+1)^2\Psi_k'(\varphi) + \Psi_k'''(\varphi)}{\lambda_k - 1},$$

where $\lambda_k = \min\{\lambda : \sin(\lambda\,\omega_k) + \lambda\sin\omega_k = 0 \text{ and } \lambda > 0\}$.

Thus, for the corner $\omega_1 = \frac{3\pi}{2}$, we have $\lambda_1 \approx 0.544483$, for $\omega_2 = \frac{5\pi}{4}$, $\lambda_2 \approx 0.673583$, and for $\omega_3 = \frac{9\pi}{8}$, $\lambda_3 \approx 0.800766$. The proposed solution is analytical in $\bar{\Omega}_k \setminus (0,0)$, but unfortunately, $P \notin W_2^1(\Omega_k)$, $\mathbf{u} \notin \mathbf{W}_2^2(\Omega_k)$.

In numerical experiments, we use meshes with a various step size h and number N, where $N \cdot h$ equals two. The approximate generalized solution (velocity field) by classical FEM converges to the exact one in the $\mathbf{W}_2^1(\Omega_k)$ norm with a rate depending on the size of reentrant corner ω, the so-called pollution effect (see [12] and the references therein): for a corner $\omega_1 = \frac{3\pi}{2}$, we have the rate of convergence, which is equal to $\mathcal{O}(h^{0.55})$, for a corner $\omega_2 = \frac{5\pi}{4}$, $\mathcal{O}(h^{0.67})$, and for a corner $\omega_3 = \frac{9\pi}{8}$, $\mathcal{O}(h^{0.8})$ (see Table 1); whereas, the approximate R_ν-generalized solution by the presented weighted FEM converges to the exact one in the $\mathbf{W}_{2,\nu}^1(\Omega_k)$ norm with a rate that is independent of the value of the internal angle ω and has the first order by h (see Table 2), where we derive computationally the optimal parameters δ, $\nu^\diamond = \nu_{opt}^\diamond$ and ν. Both errors for the R_ν-generalized and generalized solutions visually are represented in Figure 3 for different values of a number N.

Table 1. The generalized solution error $(\mathbf{u}^h - \mathbf{u})$ in the norm of a space $\mathbf{W}_2^1(\Omega_k)$.

ω_k, $N =$	74	148	296
$\frac{3\pi}{2}$	2.886×10^{-1}	1.980×10^{-1}	1.358×10^{-1}
$\frac{5\pi}{4}$	1.622×10^{-1}	1.017×10^{-1}	6.377×10^{-2}
$\frac{9\pi}{8}$	6.747×10^{-2}	3.870×10^{-2}	2.220×10^{-2}

Table 2. The R_ν-generalized solution error $(\mathbf{u}_\nu^h - \mathbf{u}_\nu)$ in the norm of a space $\mathbf{W}_{2,\nu}^1(\Omega_k)$, where $\nu^\diamond = \mu^\diamond = \lambda_k - 1$ and $\nu^\diamond = \mu^\diamond = \nu_{opt}^\diamond$.

			$\nu^\diamond = \mu^\diamond = \lambda_k - 1$			$\nu^\diamond = \mu^\diamond = \nu_{opt}^\diamond$		
ω_k	ν	δ, $N =$	74	148	296	74	148	296
$\frac{3\pi}{2}$	1.6	0.01375	2.261×10^{-4}	1.126×10^{-4}	5.504×10^{-5}	1.614×10^{-5}	8.026×10^{-5}	3.991×10^{-5}
		0.01625	3.181×10^{-4}	1.582×10^{-4}	7.895×10^{-5}	2.290×10^{-4}	1.138×10^{-4}	5.648×10^{-5}
	1.9	0.01375	6.236×10^{-5}	3.101×10^{-5}	1.543×10^{-5}	4.469×10^{-5}	2.235×10^{-5}	1.109×10^{-5}
		0.01625	9.311×10^{-5}	4.601×10^{-5}	2.288×10^{-5}	6.789×10^{-5}	3.381×10^{-5}	1.675×10^{-5}
$\frac{5\pi}{4}$	1.6	0.01375	1.181×10^{-4}	5.849×10^{-5}	2.925×10^{-5}	9.247×10^{-5}	4.603×10^{-5}	2.276×10^{-5}
		0.01625	1.720×10^{-4}	8.568×10^{-5}	4.275×10^{-5}	1.322×10^{-4}	6.567×10^{-5}	3.260×10^{-5}
	1.9	0.01375	3.320×10^{-5}	1.651×10^{-5}	8.234×10^{-6}	2.605×10^{-5}	1.293×10^{-5}	6.437×10^{-6}
		0.01625	5.115×10^{-5}	2.547×10^{-5}	1.262×10^{-5}	3.835×10^{-5}	1.905×10^{-5}	9.513×10^{-6}
$\frac{9\pi}{8}$	1.6	0.01375	6.020×10^{-5}	2.993×10^{-5}	1.495×10^{-5}	4.493×10^{-5}	2.233×10^{-5}	1.104×10^{-5}
		0.01625	7.947×10^{-5}	3.946×10^{-5}	1.959×10^{-5}	6.124×10^{-5}	3.036×10^{-5}	1.497×10^{-5}
	1.9	0.01375	1.684×10^{-5}	8.366×10^{-6}	4.170×10^{-6}	1.239×10^{-5}	6.158×10^{-6}	3.068×10^{-6}
		0.01625	2.364×10^{-5}	1.174×10^{-5}	5.800×10^{-6}	1.756×10^{-5}	8.708×10^{-6}	4.324×10^{-6}

Let $\delta'_{ji} = |u_j(M_i) - u_j^h(M_i)|$ and $\delta_{ji} = |u_j(M_i) - u_{\nu,j}^h(M_i)|$, $j = 1,2$, $M_i \in G_\Omega$ be errors for the generalized and R_ν-generalized solutions, respectively. Then, we show the percentage of nodes, where

δ'_{1i} and δ_{1i} are less than a given value $\bar{\triangle}_l$. The quantity of points $M_i \in G_\Omega$, where $\delta'_{1j} < \bar{\triangle}_l$ (for the classical FEM), is significantly less in relation to the quantity of points $M_i \in G_\Omega$, where $\delta_{1j} < \bar{\triangle}_l$ (for the proposed weighted FEM) for all sizes of the reentrant corner ω (see Table 3). Moreover, in numerical experiments, the number of nodes M_i, where $\delta'_{2i} < \bar{\triangle}_l$ and $\delta_{2i} < \bar{\triangle}_l$ are approximately equal to the number of nodes M_i, where $\delta'_{1i} < \bar{\triangle}_l$ and $\delta_{1i} < \bar{\triangle}_l, l = 1, 2$, respectively.

Table 3. The percentage of points $M_i \in G_\Omega$, where the values δ_{1i} and δ'_{1i} are less than $\bar{\triangle}_l, l = 1, 2$.

		R_ν-Generalized Solution, $\nu = 1.9$, $\delta = 0.01375, \nu^\diamond = \mu^\diamond = \nu^\diamond_{opt}$			Generalized Solution		
ω_k	$\bar{\triangle}_l, N =$	74	148	296	74	148	296
$\frac{3\pi}{2}$	10^{-5}	19.1%	36.7%	65.7%	13.2%	14.8%	22.1%
	5×10^{-6}	16.4%	29.3%	51.4%	6.2%	9.1%	15.6%
$\frac{5\pi}{4}$	10^{-5}	33.9%	51.0%	76.2%	21.4%	32.4%	44.2%
	5×10^{-6}	24.1%	42.4%	64.7%	11.4%	17.5%	27.4%
$\frac{9\pi}{8}$	10^{-5}	60.3%	91.5%	98.1%	44.7%	68.3%	86.4%
	5×10^{-6}	39.7%	62.7%	80.5%	24.8%	32.7%	44.3%

Then, we present the distribution of errors δ_{ji} and δ'_{ji} in the points M_k for components $u^h_{\nu,j}$ and u^h_j for all sizes $\omega_l, l = 1, 2, 3, j = 1, 2$, and h, such that $N = 148$ and $N = 296$. The weighted finite element method allows us to perform computations with high accuracy both inside of the domain and near the point of singularity. Moreover, the error of the proposed FEM is localized near the point of singularity and does not extend into the interior of the domain, in contrast to the error of the classical FEM for all values of the internal corner ω (see Figures 4–15).

In Figures 16–18, we show the dependence of error in the $\mathbf{W}^1_{2,\nu}(\Omega_k)$ norm on the parameter $\nu^\diamond$ ($\mu^\diamond = \nu^\diamond$), where each minimum is compatible with the best value $\nu^\diamond_{opt}$. Any value from the interval $(\lambda_k - 1, 0)$ can be taken as an exponent $\nu^\diamond$ for the presented FEM in the domain Ω_k with a reentrant corner ω_k. Moreover, if the exponent $\mu^\diamond$ does not coincide with $\nu^\diamond$, then we get substantially worse results. This research was supported in through computational research provided by the Shared Facility Center "Data Center of FEB RAS".

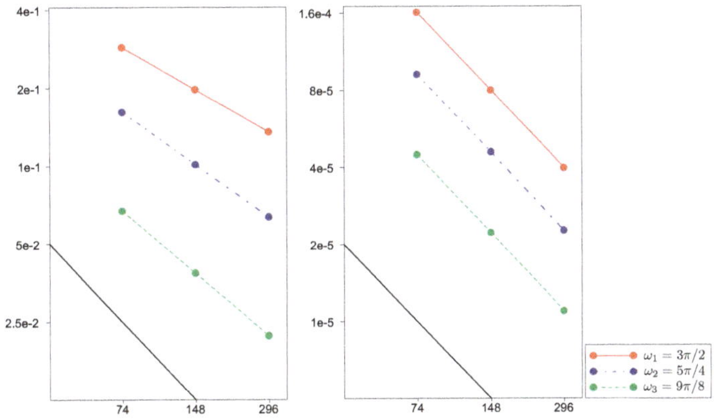

Figure 3. The errors of (**left**) a classical FEM in the $\mathbf{W}^1_2$ norm and (**right**) a weighted FEM in the $\mathbf{W}^1_{2,\nu}$ norm, where $\nu = 1.6, \delta = 0.01375 : \omega_1 = \frac{3\pi}{2}, \nu^\diamond = \nu_{opt} = -0.35; \omega_2 = \frac{5\pi}{4}, \nu^\diamond = \nu_{opt} = -0.25; \omega_3 = \frac{9\pi}{8}, \nu^\diamond = \nu_{opt} = -0.125$, for different values of a number N.

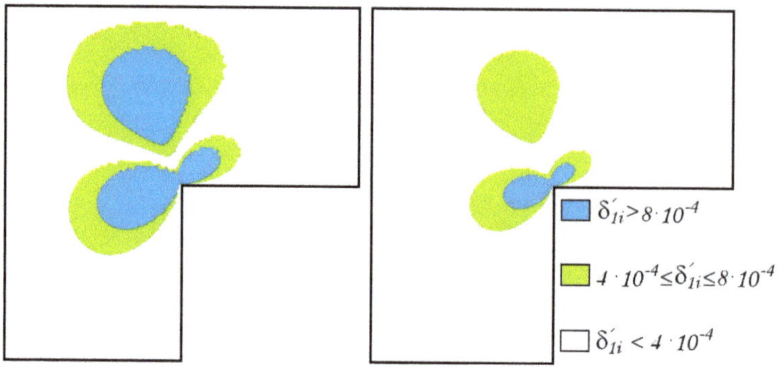

Figure 4. The errors δ'_{1i} of the approximate generalized solution (u_1^h) : $\omega_1 = \frac{3\pi}{2}$, (**left**) $N = 148$, (**right**) $N = 296$.

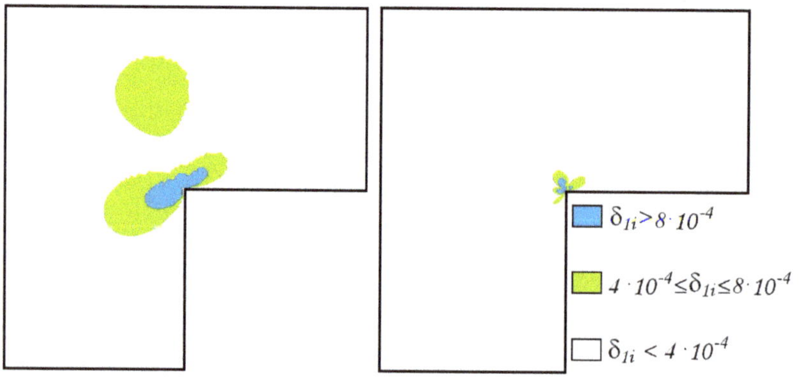

Figure 5. The errors δ_{1i} of the approximate R_ν-generalized solution $(u_{\nu,1}^h)$: $\omega_1 = \frac{3\pi}{2}$, $\nu = 1.6$, $\delta = 0.01375$, $\nu^\diamond = \mu^\diamond = -0.35$, (**left**) $N = 148$, (**right**) $N = 296$.

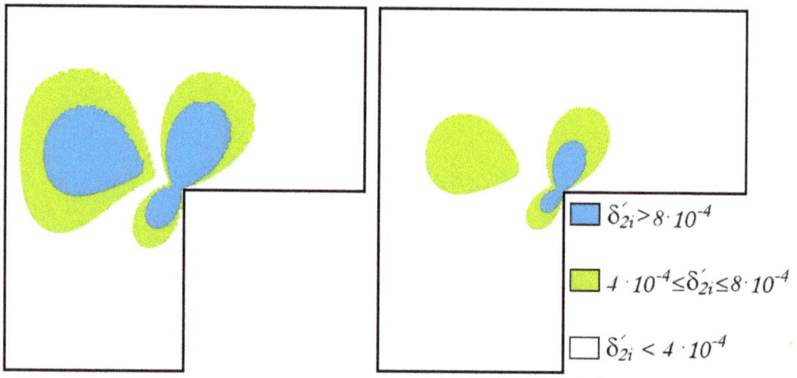

Figure 6. The distribution of the errors δ'_{2i} of the approximate generalized solution (u_2^h) : $\omega_1 = \frac{3\pi}{2}$, (**left**) $N = 148$, (**right**) $N = 296$.

Figure 7. The errors δ_{2i} of the approximate R_ν-generalized solution $(u^h_{\nu,2})$: $\omega_1 = \frac{3\pi}{2}$, $\nu = 1.6$, $\delta = 0.01375$, $\nu^\diamond = \mu^\diamond = -0.35$, **(left)** $N = 148$, **(right)** $N = 296$.

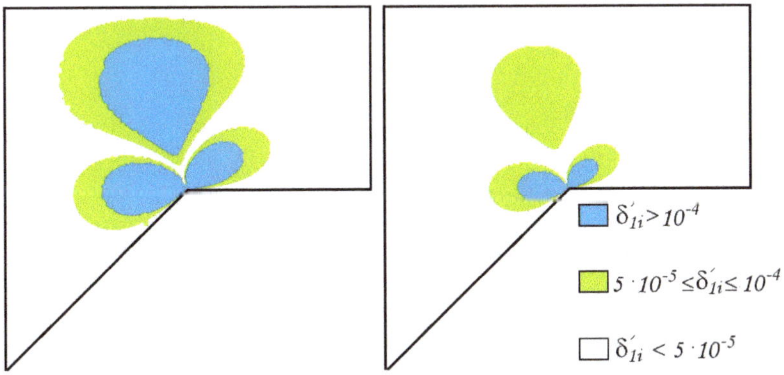

Figure 8. The errors δ'_{1i} of the approximate generalized solution (u^h_1) : $\omega_2 = \frac{5\pi}{4}$, **(left)** $N = 148$, **(right)** $N = 296$.

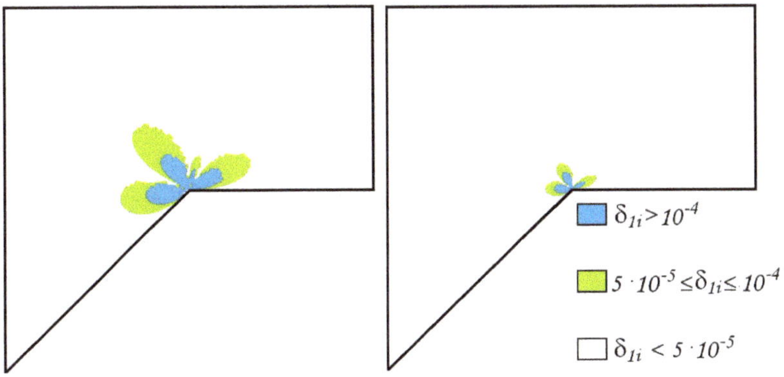

Figure 9. The errors δ_{1i} of the approximate R_ν-generalized solution $(u^h_{\nu,1})$: $\omega_2 = \frac{5\pi}{4}$, $\nu = 1.6$, $\delta = 0.01375$, $\nu^\diamond = \mu^\diamond = -0.25$, **(left)** $N = 148$, **(right)** $N = 296$.

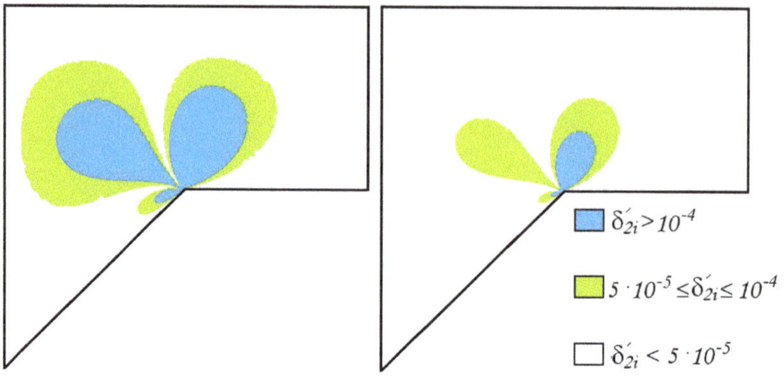

Figure 10. The errors δ'_{2i} of the approximate generalized solution (u_2^h) : $\omega_2 = \frac{5\pi}{4}$, **(left)** $N = 148$, **(right)** $N = 296$.

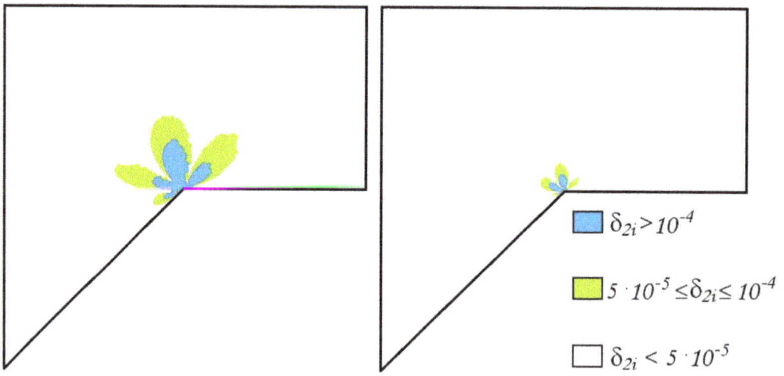

Figure 11. The errors δ_{2i} of the approximate R_ν-generalized solution $(u_{\nu,2}^h)$: $\omega_2 = \frac{5\pi}{4}$, $\nu = 1.6$, $\delta = 0.01375$, $\nu^\diamond = \mu^\diamond = -0.25$, **(left)** $N = 148$, **(right)** $N = 296$.

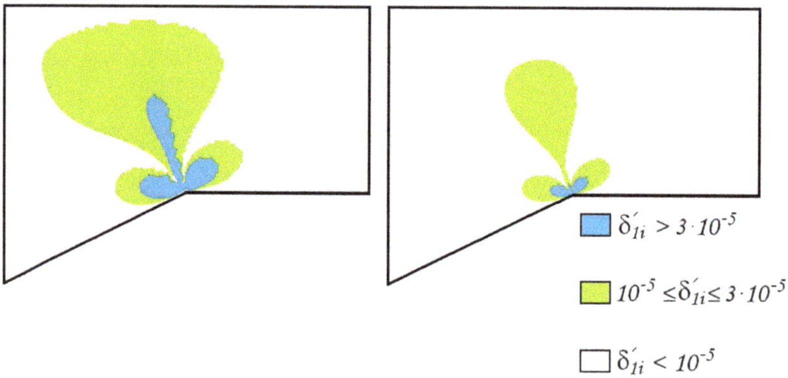

Figure 12. The errors δ'_{1i} of the approximate generalized solution (u_1^h) : $\omega_3 = \frac{9\pi}{8}$, **(left)** $N = 148$, **(right)** $N = 296$.

Figure 13. The errors δ_{1i} of the approximate R_ν-generalized solution $(u^h_{\nu,1})$: $\omega_3 = \frac{9\pi}{8}$, $\nu = 1.6$, $\delta = 0.01375$, $\nu^\diamond = \mu^\diamond = -0.125$, **(left)** $N = 148$, **(right)** $N = 296$.

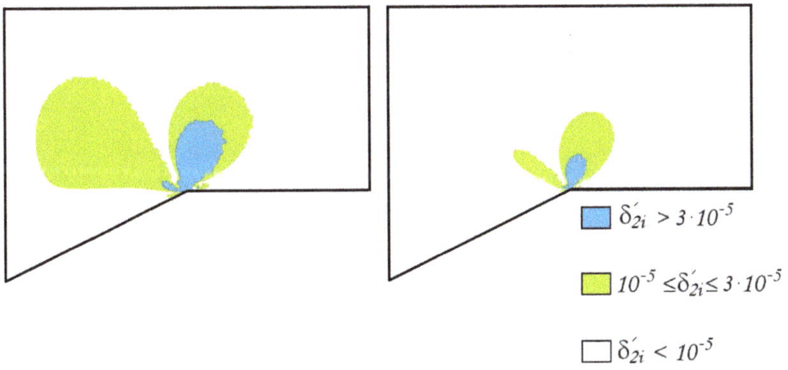

Figure 14. The errors δ'_{2i} of the approximate generalized solution (u^h_2) : $\omega_3 = \frac{9\pi}{8}$, **(left)** $N = 148$, **(right)** $N = 296$.

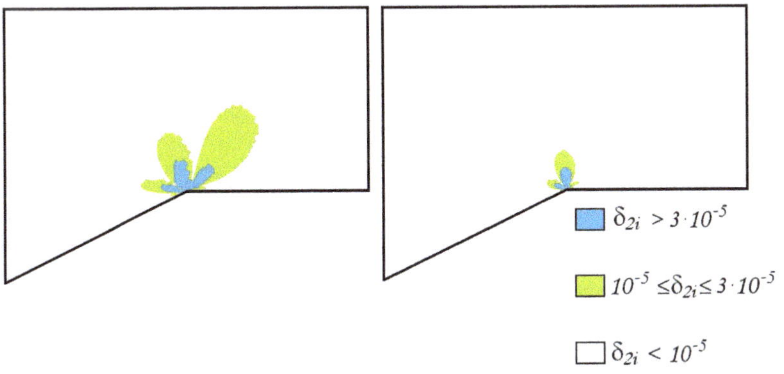

Figure 15. The errors δ_{2i} of the approximate R_ν-generalized solution $(u^h_{\nu,2})$: $\omega_3 = \frac{9\pi}{8}$, $\nu = 1.6$, $\delta = 0.01375$, $\nu^\diamond = \mu^\diamond = -0.125$, **(left)** $N = 148$, **(right)** $N = 296$.

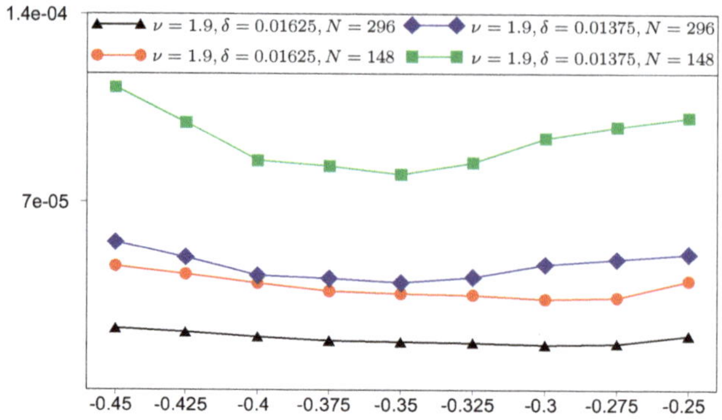

Figure 16. The dependence of error $(\mathbf{u}_\nu^h - \mathbf{u})$ in the $\mathbf{W}^1_{2,\nu}(\Omega_1)$ norm on the degree $\nu^\diamond$, $\omega_1 = \frac{3\pi}{2}$.

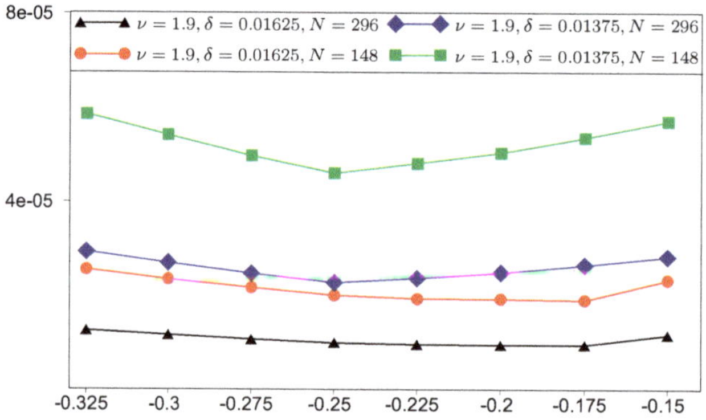

Figure 17. The dependence of error $(\mathbf{u}_\nu^h - \mathbf{u})$ in the $\mathbf{W}^1_{2,\nu}(\Omega_2)$ norm on the degree $\nu^\diamond$, $\omega_2 = \frac{5\pi}{4}$.

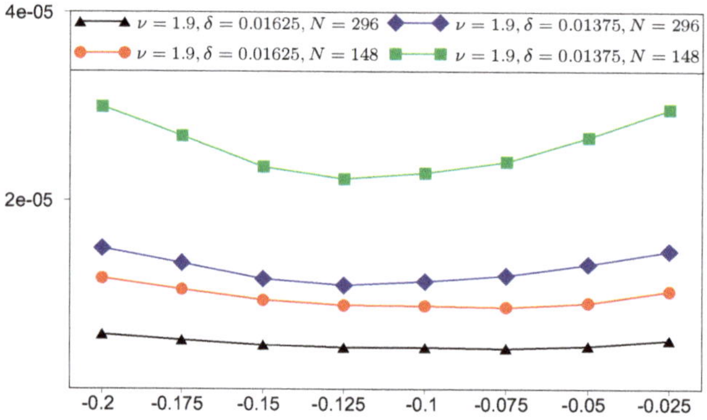

Figure 18. The dependence of error $(\mathbf{u}_\nu^h - \mathbf{u})$ in the $\mathbf{W}^1_{2,\nu}(\Omega_3)$ norm on the degree $\nu^\diamond$, $\omega_3 = \frac{9\pi}{8}$.

6. Conclusions

The main results of the numerical experiments for the Oseen problem (8) and (9) lead to the following conclusions:

- The approximate generalized solution (velocity field) by classical FEM converges to the exact one in the $\mathbf{W}_2^1(\Omega_k)$ norm with a rate $\mathcal{O}(h^{\lambda}), \lambda < 1$, where the exponent λ depends on the size of reentrant corner ω, the so-called pollution effect (see [12] and the references therein), while the approximate R_ν-generalized solution by the presented weighted FEM converges to the exact one in the $\mathbf{W}_{2,\nu}^1$ norm with a rate that is independent of the value of the internal angle ω and has the first order by h for various values of ν, δ (see Tables 1 and 2 and Figure 3).
- Thanks to Theorem 3.1 in [13], there exists a limitation on the radius δ_* of the neighborhood of a reentrant corner ω and $\rho(\mathbf{x})$ exponent ν_* in Definition 1, that for all $\delta < \delta_*$ and $\nu > \nu_*$, a weighted inf-sup condition holds. After a series of computational experiments, we conclude that $\nu_* \sim 1$ and $\delta_* \sim h$.
- The proposed approach allows us to compute the approximate solution by the weighted FEM with a given accuracy 10^{-3}, for example in a case when the internal corner ω is equal to $\frac{3\pi}{2}$, about 10^6-times faster than using classical FEM. Note that in implementing the weighted FEM, one can spend about 10^6-times less computing resources and energy consumption.
- The weighted finite element method enables us to perform computations with high accuracy, both inside of the domain and near the point of singularity.

Author Contributions: V.A.R. and A.V.R. contributed equally in each stage of the work. All authors read and approved the final version of the paper.

Funding: The reported study was supported by RFBR and RSF according to the research project Nos. 19-01-00007-a and 19-71-20006, respectively.

Acknowledgments: We would like to thank the Philip Li, Elaine Chen and the referees for their invaluable suggestions due to which the manuscript was significantly improved.

Conflicts of Interest: The authors declare no conflict of interest.

References

1. Ciarlet, P. *The Finite Element Method for Elliptic Problems*; Studies in Mathematics and Its Applications; North-Holland: Amsterdam, The Netherlands, 1978.
2. Rukavishnikov, V.A. On the differential properties of R_ν-generalized solution of Dirichlet problem. *Dokl. Akad. Nauk SSSR* **1989**, *309*, 1318–1320.
3. Rukavishnikov, V.A.; Rukavishnikova, H.I. The finite element method for a boundary value problem with strong singularity. *J. Comput. Appl. Math.* **2010**, *234*, 2870–2882. [CrossRef]
4. Rukavishnikov, V.A.; Mosolapov, A.O. New numerical method for solving time-harmonic Maxwell equations with strong singularity. *J. Comput. Phys.* **2012**, *231*, 2438–2448. [CrossRef]
5. Rukavishnikov, V.A.; Rukavishnikova, H.I. On the error estimation of the finite element method for the boundary value problems with singularity in the Lebesgue weighted space. *Numer. Funct. Anal. Optim.* **2013**, *34*, 1328–1347. [CrossRef]
6. Rukavishnikov, V.A. *Weighted FEM for Two-Dimensional Elasticity Problem with Corner Singularity*; Lecture Notes in Computational Science and Engineering; Springer Verlag; Oxford University Press: Oxford, UK, 2016; Volume 112, pp. 411–419. [CrossRef]
7. Rukavishnikov, V.A.; Rukavishnikova, H.I. *Weighted Finite-Element Method for Elasticity Problems with Singularity*; Finite Element Method Simulation, Numerical Analysis and Solution Techniques; IntechOpen Limited: London, UK, 2018; pp. 295–311. [CrossRef]
8. Benzi, M.; Golub, G.H.; Liesen, J. Numerical solution of saddle point problems. *Acta Numer.* **2005**, *14*, 1–137. [CrossRef]
9. Layton, W.; Manica, C.; Neda, M.; Olshanskii, M.; Rebholz, L.G. On the accuracy of the rotation form in simulations of the Navier-Stokes equations. *J. Computat. Phys.* **2009**, *228*, 3433–3447. [CrossRef]
10. Moffatt, H.K. Viscous and resistive eddies near a sharp corner. *J. Fluid Mech.* **1964**, *18*, 1–18. [CrossRef]
11. Dauge, M. Stationary Stokes and Navier-Stokes system on two- or three-dimensional domains with corners. I. Linearized equations. *SIAM J. Math. Anal.* **1989**, *20*, 74–97. [CrossRef]
12. Blum, H. The influence of reentrant corners in the numerical approximation of viscous flow problems. In *Numerical Treatment of the Navier-Stokes Equations*; Springer: Berlin, Germany, 1990; Volume 30. [CrossRef]

13. Rukavishnikov, V.A.; Rukavishnikov, A.V. Weighted finite element method for the Stokes problem with corner singularity. *J. Comput. Appl. Math.* **2018**, *341*, 144–156. [CrossRef]
14. Brezzi, F.; Fortin, M. *Mixed and Hybrid Finite Element Methods*; Springer-Verlag: New York, NY, USA, 1991. [CrossRef]
15. Rukavishnikov, V.A.; Rukavishnikov, A.V. New approximate method for solving the Stokes problem in a domain with corner singularity. *Bull. South Ural State Univ. Ser. Math. Model. Program. Comput. Softw.* **2018**, *11*, 95–108. [CrossRef]
16. Scott, L.R.; Vogelius, M. Norm estimates for a maximal right inverse of the divergence operator in spaces of piecewise polynomials. *Math. Model. Numer. Anal.* **1985**, *19*, 111–143. [CrossRef]
17. Linke, A. Collision in a cross-shaped domain—A steady 2D Navier-Stokes example demonstrating the importance of mass conservation in CFD. *Comput. Methods Appl. Mech. Eng.* **2009**, *198*, 3268–3278. [CrossRef]
18. Qin, J. On the Convergence of Some Low Order Mixed Finite Element for Incompressible Fluids. Ph.D. Thesis, Pennsylvania State University, University Park, PA, USA, 1994. [CrossRef]
19. Bramble, J.H.; Pasciak, J.E.; Vassilev, A.T. Analysis of the inexact Uzawa algorithm for saddle point problems. *SIAM J. Numer. Anal.* **1997**, *34*, 1072–1092. [CrossRef]
20. Saad, Y. *Iterative Methods for Sparse Linear Systems*; Society for Industrial and Applied Mathematics: Philadelphia, PA, USA, 2003. [CrossRef]

MDPI
St. Alban-Anlage 66
4052 Basel
Switzerland
Tel. +41 61 683 77 34
Fax +41 61 302 89 18
www.mdpi.com

Symmetry Editorial Office
E-mail: symmetry@mdpi.com
www.mdpi.com/journal/symmetry

www.ingramcontent.com/pod-product-compliance
Lightning Source LLC
LaVergne TN
LVHW070545100526
838202LV00012B/384
* 9 7 8 3 0 3 9 3 6 0 2 0 8 *